T0261251

**Chemical and Biomedical Engineering
Calculations Using Python**®

Chemical and Biomedical Engineering Calculations Using Python®

Jeffrey J. Heys
Montana State University
Bozeman, Montana, USA

This edition first published 2017
© 2017 John Wiley & Sons, Inc

The right of Jeffrey J. Heys to be identified as the author of this work has been asserted in accordance with law.

Registered Office
John Wiley & Sons, Inc., 111 River Street, Hoboken, NJ 07030, USA

Editorial Office
111 River Street, Hoboken, NJ 07030, USA

For details of our global editorial offices, customer services, and more information about Wiley products visit us at www.wiley.com.

Wiley also publishes its books in a variety of electronic formats and by print-on-demand. Some content that appears in standard print versions of this book may not be available in other formats.

Library of Congress Cataloging-in-Publication Data

Names: Heys, Jeffrey J., 1974- author.
Title: Chemical and Biomedical Engineering Calculations Using Python®/ Jeffrey J. Heys.
Description: Hoboken, NJ : John Wiley & Sons, 2017. | Includes
  bibliographical references and index.
Identifiers: LCCN 2016039763| ISBN 9781119267065 (cloth) | ISBN 9781119267072
  (epub)
Subjects: LCSH: Engineering mathematics. | Python (Computer program language)
Classification: LCC TA330 .H49 2017 | DDC 620.00285/5133–dc23 LC record available at
  https://lccn.loc.gov/2016039763

Cover design by Wiley

Cover image: © lvcandy/Getty Images, Inc.

Set in 10/12pt, Warnock by SPi Global, Chennai, India

Printed in United States

10 9 8 7 6 5 4 3 2 1

Contents

Preface

Computers have become a powerful tool in the field of engineering. Before the widespread availability of computers, mathematical models of engineering problems needed to be simplified to the point that the calculations could be reliably performed by a single individual using a calculator or slide rule, and, fortunately, for many engineering problems, simplified models were adequate. However, as process complexity and engineering design complexity increased, engineers increasingly turned to computers for help in managing and automating the large number of calculations required.

The computational tools used by engineers have evolved considerably over the past few decades. In the 1960s and 1970s, computers were not widely available, and they were a specialized tool that was operated by highly trained individuals. In the 1980s and 1990s, computers became widely available, but the engineering software and computational tools were relatively simple compared to what is available in the twenty-first century. The individual that was using the computer general understood the calculations that were being performed, and the computer was primarily a tool for automating those calculations. Many engineering students during this time learned to program in either FORTRAN or C, and the programs written by engineers were frequently limited to a few hundred lines of code. More specialized and easier to use programming environments like MATLAB and IDL were also developed during the 1980s, and they usually helped to decrease the time required to write a computer algorithm, but they increased the time required to execute or run the algorithm.

The trend toward greater specialization and ease of use in computational tools continued in the twenty-first century. The various fields of engineering saw an exponential increase in powerful and easy-to-use tools like AutoCAD, SolidWorks, ANSYS, and Aspen. (Clearly, it is a good idea to choose a name for your software that begins with "A" so it appears first alphabetically.) The individual that uses these software packages may have some understanding of the calculations that are being performed, but they almost never fully understand the calculations and in some cases have no understanding of the mathematics that is being performed by the computer. Today, engineering students are typically taught to use multiple computational software packages during the typical undergraduate education. The irony of this situation is that students often do not understand the calculation being performed by the software – they do not know the limitations of the mathematical models, they do not know the expected accuracy of the approximate solution, and they do not always have the intuition necessary to recognize a highly incorrect result. Another loss associated with the rise of specialized software tools for engineers is that it is often very difficult to find a computational tool for a new problem. The software often works well for the limited range of problems for which it was designed, but, if an engineer wishes to analyze something new or include some change that takes the problem just beyond the range of problems for which the software was design, that engineer is often "out of luck" because no computational tool is available to help.

I do not advocate abandoning modern engineering software. I do not advocate returning to the use of custom FORTRAN computer codes for every problem. I do advocate that engineering students get some experience writing short computer programs. This experience teaches one to think precisely as computers are notoriously unforgiving when we make mistakes in our logic. It teaches one to decompose a complex process down into small, individual steps. This experience teaches one to develop a unique solution for a new problem that is not handled well by existing software. Finally, the experience of creating a computer algorithm helps to develop a recognition of when computations are likely to be reliable and when they are not – when the computational solution is sufficiently accurate and when it is not.

The goal of this book is to provide the reader with an understanding of standard computational methods for approximating the solution to common problems in Chemical and Biomedical Engineering. The book does not have a comprehensive coverage of computational methods, but it is instead intended to provide the introductory coverage necessary to understand the most commonly used algorithms. The computer language used to explore the different computational methods is Python. The advantages of using Python include its wide and growing popularity, large library of existing algorithms, and its licensing as free, open source software. The final and possibly greatest advantage in using Python is that it is easy to learn to write general computational algorithms and more specialized numerical algorithms are also easy to write, thanks to the NumPy and SciPy libraries. By the end of this book, the reader should have a solid understanding of how to write and use computational algorithms in Python to solve common mathematical problems in Chemical and Biomedical Engineering.

The course that motivated the creation of this textbook is one semester of approximately 15 weeks. It is my belief that most of this material can be covered in that length of time. Each chapter in the textbook covers a different topic and the book was constructed so that the material in that chapter could be covered in approximately 1 week. There are, of course, some exceptions. The large number of topics and short amount of time associated with a single semester may encourage instructors using this book to consider a slightly different format than the traditional lecture format. For example, if two class times per week are available, an instructor may want to consider requiring students to read the book or watch an online lecture that presents the material to be covered before coming to the first class meeting time each week. The two class periods could then be used to cover example problems (the first class each week) and a "working class" could be used for the second class meeting of the week. When students are trying to complete the homework, they often need support to overcome a difficult error message or unexpected and unphysical numerical answer from the computer, and allowing students to work on problems for one class time per week is often very beneficial.

Suggested homework problems are included at the end of each chapter. Many of the homework problems are written so that the person answering the problem must respond to a request from a real or hypothetical organization such as a company or government agency. The author of this book typically assigns one or two problems per week and requires students to submit their solutions in the form of a memo to the organization that posed the problem. The memo typically is about 1 page of text plus 1–3 figures for a total of 2 or 3 pages for the main body of the memo, and the Python code is included by the student in an appendix with the memo. Requiring students to practice technical writing is a benefit of using this approach, and many students are motivated when the problems have more of a "real world" flavor and are less abstract.

In closing, I would like to offer my sincerest thanks and gratitude to the many unnamed individuals that have contributed to building Python and making scientific computing using Python such a wonderful reality. To me, it is really humbling and encouraging to see the great work that these individuals have freely given to the world. I would like to single out two individuals by name because of the transformative impact of their work – without their work, I would never have started using Python as extensively as I do, and this book would never have been written. The first individual is Travis Oliphant, the primary creator of NumPy and the founder of Continuum Analytics, which produces the Anaconda Python Distribution. The second individual is Fernando Perez, a physicist, creator of iPython, and, most importantly to me, the person that came into my office at the University of Colorado at Boulder and told me that I should try learning Python because it made programming fun!

Bozeman, Montana   *Jeffrey J. Heys*
July 15, 2016

About the Companion Website

This book is accompanied by a companion website:

www.wiley.com/go/heys/engineeringcalculations_python

The website includes:

- Python Computer Codes.

1

Problem Solving in Engineering

In chemical and biological engineering, students find that the sequence of steps outlined in Figure 1.1 is an effective problem-solving approach for the vast majority of the problems they encounter.

In most courses, students practice all the steps outlined in Figure 1.1, but the focus is usually on the construction of the system diagram and developing the mathematical equations for every unique type of process that is described in a particular course. Only limited attention is usually given to solving the mathematical equations that arise in a particular course because the assumption is that the student should have learned how to do that in their mathematics courses or some other course. Many engineering curricula have a course that is focused on the use of computers to solve the many different types of equations that arise in a student's engineering courses. The focus of this textbook is just "using computers to solve the equation(s) that students typically encounter throughout the engineering curriculum."

The timing of a course on computational or numerical methods for solving engineering problems varies considerably from one curriculum to the next. One approach is to schedule the course near the end of the curriculum. As an upper level course, students are able to review most of the engineering principles and mathematics that they learned previously and develop a new set of tools (specifically, computational tools) for solving those same problems. Two disadvantages are associated with this approach. First, students do not have the computational tools when they first learn a new engineering principle, which limits the scope of problems they can solve to problems that can be largely solved without a computer (i.e., problems that can be solved with paper and pencil). The second disadvantage is that the third and fourth years of many engineering curricula are already filled with other required courses and it is difficult to find time for yet another course.

A second approach is to schedule the computational methods course early in the curriculum, before students have taken most of the engineering courses in which they learn to derive, construct, and identify the mathematical equations they need to solve and that sometime require a computational approach. There are also two problems with this approach. First, the students have

Chemical and Biomedical Engineering Calculations Using Python®, First Edition. Jeffrey J. Heys.
© 2017 John Wiley & Sons, Inc. Published 2017 by John Wiley & Sons, Inc.
Companion Website: www.wiley.com/go/heys/engineeringcalculations_python

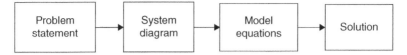

Figure 1.1 Engineering problem-solving process.

typically not taken all the required mathematics courses, and, as a result, it is difficult to teach a computational approach to solving a differential equation when a student is not yet familiar with differential equations or techniques for solving them. The second disadvantage is that the student has not taken courses on separations, kinetics, transport, and so on in which they learn to derive or identify the appropriate mathematical equation(s) for their particular problem. It is, of course, difficult to teach a computational approach to solving an equation when the importance or relevance of that equation is not known.

A third approach for addressing this dilemma is to simply not teach a stand-alone computational methods course and instead cover the relevant computational approaches as they are needed in each individual course. We will continue our listing of the "top two challenges" and identify two potential difficulties with this approach. First, instead of learning and becoming comfortable with two or three computational tools (i.e., mathematical software packages), students under this format often need to learn 4 or 5 computational tools because every one of their instructors prefers a different tool, and the students never really become proficient with any single tool. The second difficulty is that there are a few important concepts that play a role in many of the various computational methods, for example, rounding error, logical operators, and accuracy, that may never be taught if there is not a single course focused on computational methods.

This textbook, and the course that it was originally written to support, is focused on the second approach – a course that appears in the first year or early in the second year of an engineering curriculum. The main reason for adopting this approach is simply the belief that it is critical for students to understand both the potential power and flexibility of computational methods and also the important limitations of these methods before using them to solve problems in engineering. For a student to use a computational tool in a course and blindly trust that tool because they do not understand the algorithms behind the tool is probably more destructive than never learning the tool at all. Further, to limit a student to only problems that can be solved with paper and pencil for most of their undergraduate education is similarly unacceptable. Addressing the limitations associated with teaching computational methods before most of the fundamental engineering and some mathematics courses is difficult. The basic strategy employed by this book is to teach students to recognize the type of mathematical equation they need to solve, and, once they know the type of equation, they can take advantage of the appropriate computational approach that is presented here (or, more likely, refer back to this book for the appropriate algorithm for their particular equation).

There is a second, and possibly more important, reason for learning this material early in the engineering education process. It is related to the fact that one of the most difficult skills for many science, engineering, and mathematics students to master is the ability to combine a number of small, simple pieces together into a more complex framework. In most science, engineering, and mathematics courses in high school and early in college, students learn to find the right equation to solve the question they are asked to answer. Most problems can be completed in *one or two* steps. Problems in later courses, on the other hand, can often require *5–10* or more steps and can require multiple pages of equations and mathematics to solve. This transition from small problems that only require a few lines to large problems that require a few pages can be very challenging for many science, engineering, and mathematics students. I believe that programming in general, and numerical computations, in particular, can be a great way to develop the skills associated with solving larger problems. Programming requires one to combine a number of simple logical commands and variables together into a more complex framework. Programming develops the parts of our brains that allow us to synthesize a number of smaller pieces into a much larger whole. A good analogy is building something complex (e.g., the Death Star) with LEGO bricks. This process requires one to properly and carefully combine a number of simple pieces into a much larger structure. The entire process requires one to simultaneously think on both the large scale ("What is my design objective?") and the small scale ("Will these two pieces stay connected? Are they compatible?"). This skill is necessary for both programming and engineering. It is a skill that almost everyone is capable of developing, but it takes practice – so, we might as well start early!

This textbook advocates that students develop the following skills: (1) recognize the type of mathematical equation that needs to be solved – algebraic or differential? linear or nonlinear? interpolation or regression? ordinary or partial differential equation (PDE)?, and (2) select and implement the appropriate algorithm. If students are able to develop these two skills, they will be equipped with a set of tools that will serve them well in their later engineering courses. These tools can be used by a student to check their work, even when they are primarily using paper and pencil to solve a problem. It is not optimal that students learn how to approximately solve mathematical equations before they know why the equation is relevant, but every effort is made in this book to at least try and explain the relevance of equations when possible.

1.1 Equation Identification and Categorization

We identified two categories of skills that we wish to develop throughout this book: (1) recognizing the type of mathematical equation(s) and (2) selecting and implementing an appropriate computational method. The first skill will be covered in this chapter and then the remainder of the book is for developing the second set of skills.

1.1.1 Algebraic versus Differential Equations

The distinction between algebraic and differential equations is trivial – a differential equation is a relationship between the derivatives of a variable and some function. Differential equations described the rate of change of a variable; typically the rate of change with respect to space or time. Equations can have both independent and dependent variables. It is usually simplest to identify the dependent variables because their value depends on the value of another variable. For example, in both $v(t) = 2\pi + t^2$ and $\frac{dv}{dt} = 3 + v \cdot t$, v is the dependent variable because its value depends on the value of t and t is the independent variable. There can be multiple independent variables, for example, multiple spatial dimensions and time, and the value of dependent variable may depended on the value of all independent variables. The density of air, for example, varies with location: latitude, longitude, and elevation above sea level, as well as time. Therefore, if we have an equation that describes the density of air as a function of location and time, then, in that equation, density is the dependent variable and location and time are the independent variables. Similarly, the ideal gas law can be used to calculate the density of air: $\rho(P, T) = \frac{P}{R \cdot T}$. For this equation, ρ is a function of temperature and pressure, so ρ is the dependent variable and P and T are the independent variables. Alternatively, this equation could be seen such that pressure, P, is the dependent variable that depends on density, ρ, and temperature, T, that is, $P(\rho, T) = \rho \cdot R \cdot T$.

For differential equations, there are three different notation styles that are commonly used for derivatives.

Leibniz notation The derivative of the function, $f(x)$, with respect to x is written as

$$\frac{df}{dx}$$

and the second derivative is written:

$$\frac{d^2 f}{dx^2}.$$

The *partial* derivative of $f(x, y)$ with respect to x is

$$\frac{\partial f}{\partial x}.$$

Lagrange notation The derivative of the function, $f(x)$, with respect to x is written:

$$f'(x)$$

and the second derivative is written:

$$f''(x).$$

The notation is not easily extended to partial derivatives and there is no universal standard, but one style that is used is to switch from the prime mark, ', to a subscript so that the partial derivative of $f(x, y)$ with respect to x is

$$f_x.$$

Euler notation The derivative of the function, $f(x)$, with respect to x is written:

$$D \, f$$

and the second derivative is written:

$$D^2 \, f.$$

The *partial* derivative of $f(x, y)$ with respect to x is

$$D_x \, f.$$

In summary, differential equations have at least one derivative and algebraic equations do not. The presence of a derivative has a significant impact on the computational method used for solving the problem of interest.

1.1.2 Linear versus Nonlinear Equations

A linear function, $f(x)$, is one that satisfies both of the following properties:

additivity: $f(x + y) = f(x) + f(y)$.
homogeneity: $f(c \cdot x) = cf(x)$.

In practice, this means that the dependent variables cannot appear in polynomials of degree two or higher (i.e., $f(x) = x^2$ is nonlinear because $(x + y)^2 \neq x^2 + y^2$), in nonlinear arguments within the function (i.e., $f(x) = x + \sin(x)$ is nonlinear because $\sin(x + y) \neq \sin(x) + \sin(y)$), or as products of each other (i.e., $f(x, y) = x + xy$ is nonlinear).

For algebraic equations, it is typically straightforward to solve linear systems of equations, even very large systems consisting of millions of equations and millions of unknowns. Two different methods for solving linear systems of equations will be covered in Chapter 6. Nonlinear algebraic equations can sometimes be solved exactly using techniques learned in algebra or using symbolic mathematics algorithms, especially when there is only a single equation. However, if we have more than one nonlinear equation or even a single, particularly complex nonlinear algebraic equation (or if we are simply feeling a little lazy), we may need to take advantage of a computational technique to try and find an approximate solution. Algorithms for solving nonlinear algebraic equations are described in Chapter 8.

It is important to note that the distinction between linear and nonlinear equations can also be extended to differential equations and all of the same principles apply. For example, $\frac{dc}{dt} = 4c$ and $\frac{d^2c}{dt^2} = 2\sin(\pi t)$ are linear while $\frac{dc}{dt} = c^2$ is nonlinear. In some cases, the nonlinearity will not significantly increase the computational challenge, but, in other cases like the Navier–Stokes equations, the nonlinearity can significantly increase the difficulty in obtaining even an approximate solution.

Linear versus Nonlinear Examples

Linear:
- single linear equation: $5 \cdot x + \frac{1}{3} = x$
- linear system of equations:

$$3 \cdot x + \frac{y}{4} = 10$$

$$x = 6 \cdot y.$$

Nonlinear:
- single nonlinear equation: $5 \cdot x - \frac{1}{3} = \sqrt{x}$
- single nonlinear equation: $x^2 - 8 \cdot x - 9 = 0$
- nonlinear system of equations:

$$3 \cdot x \cdot y + \frac{y}{4} = 10$$

$$x - 6 \cdot y = 0$$

- nonlinear system of equations:

$$x + y = 4$$

$$\log(x) - 7 \cdot y = 0.$$

1.1.3 Ordinary versus Partial Differential Equations

An ordinary differential equation (ODE) has a single independent variable. For example, if a differential equation only has derivatives with respect to time, t, or a single spatial dimension, x, it is an ODE. A differential equation with two or more independent variables is a PDE. The following are examples of ODEs.

$$t \cdot \frac{dp}{dt} + \frac{d^2p}{dt^2} = \sin(t) \quad \text{(linear, second-order ODE).}$$

If you have not taken a differential equations course, this equation may look a little intimidating or confusing. To solve this equation, we need to find a function $p(t)$ where the first derivative of the function, multiplied by t, plus the second derivative of the function is equal to $\sin(t)$. If that sounds difficult, do not worry, by the end of this textbook, you will know how to get an approximate solution, that is, a numerical approximation of the function $p(t)$. It is also important to emphasize that multiplying the dependent variable p by the independent variable t did not make the equation nonlinear. A nonlinearity only arises if, for example, p is multiplied by itself.

$$\frac{dx}{dt} = x^2 + 3\cos(t) \quad \text{(nonlinear, first-order ODE)}.$$

Again, if you have not had a differential equations course, solving this equation requires finding a function $x(t)$ that has a derivative equal to $(x(t))^2$ plus $3\cos(t)$. Do not worry if that makes your head spin, we will also cover the solution of this class of problems.

Some examples of PDEs are included below.

$$\frac{\partial T}{\partial t} = \alpha \frac{\partial^2 T}{\partial x^2} \quad \text{(linear, second-order PDE)}.$$

This is an equation that describes unsteady, conductive heat transport in one spatial dimension. You could use this equation to describe, for example, the warming of the ground when the sun comes up in the morning, among many other examples. Solving this equation requires finding a function $T(x, t)$ of both time t and space x where the first derivative with respect to time is equal to α times the second derivative with respect to space.

$$m\frac{\partial m}{\partial x} + \frac{\partial m}{\partial y} = 0 \quad \text{(nonlinear, first-order PDE)}.$$

By now it is probably obvious that the standard mathematical convention is to use ∂ for derivatives in a PDE while ODEs use d. The order of the equation is determined by the order of the highest derivative.

Solving a Differential Equation

Even though you may not have taken a differential equations course, you might be able to solve a simplified version of the first ODE example. Try to solve

$$\frac{d^2p}{dt^2} = \sin(t).$$

Notice that we have eliminated the difficult term with t multiplied by the first derivative. Let us start by integrating both sides of the equation with respect to t:

$$\int \frac{d}{dt}\left(\frac{dp}{dt}\right) dt = \int \sin(t) dt.$$

(Continued)

Solving a Differential Equation (Continued)

Recalling that an integral is just an antiderivative, we get

$$\frac{dp}{dt} + c_1 = -\cos(t) + c_2.$$

The two constants of integration can simply be combined into a single constant, c_0, which can be placed on the right-hand side giving:

$$\frac{dp}{dt} = -\cos(t) + c_0.$$

Now, let us integrate both sides once more with respect to t:

$$p(t) + c_3 = -\sin(t) + c_0 t + c_4,$$

which we can simplify once again by combining the two new constants of integration to a single constant c, to give

$$p(t) = -\sin(t) + c_0 t + c.$$

In order to fully determine our unknown function $p(t)$, we need two additional conditions to solve for the value of our two remaining unknown constants, c_0 and c. Typically, this additional information would be initial conditions, that is, the value of p when $t = 0$, and the value of $\frac{dp}{dt}$ at $t = 0$.

It is always a good idea to check the solution to your problem by substituting $p(t)$ back into the original differential equation and checking to make sure that the left side (i.e., the second derivate of $p(t)$) is equal to the right-hand side.

1.1.4 Interpolation versus Regression

Within engineering, it is often necessary to obtain an equation, usually a polynomial equation, that "fits" a given set of data. If we want an equation that exactly matches the data, then we must interpolate the data so that we obtain a function (e.g., a polynomial) that has the same value as the data for a given value of the independent variable (Figure 1.2). In order to determine an interpolant, the number of adjustable parameters that we determine in the equation must equal the number of data points. For example, if we want to interpolate three data points, we must use an equation that has three adjustable parameters, such as a quadratic polynomial, $ax^2 + bx + c$.

In practice, it is actually pretty rare that we want to exactly interpolate a given set of data because we hopefully have a large amount of data (and we do not want to use a very high-order polynomial) and that data contains some amount of error. In most cases, we want to approximately fit our data with an equation of some form (Figure 1.3). In order to do this, we must first decide

Figure 1.2 An example of interpolation for a set of data. The data is usually represented using points (circles) and the interpolant function is usually represented using a line.

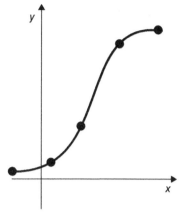

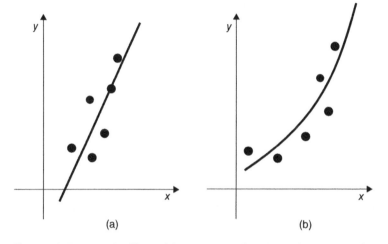

(a)   (b)

Figure 1.3 An example of linear (a) regression and nonlinear (b) regression for a set of data.

how we want to measure the "goodness" of a fit. Maybe we want to fit an equation so that the sum of the distances from the best fit equation to each and every point is minimized. Another option (the option that is almost always selected) is to minimize the sum of the *square* of the distance between every data point and the "best" fit approximation. This is the so-called least-squares regression approach. The function that gives us the best fit based on our chosen criteria is called the regression function and the process of determining the regression function is called regression analysis. The most popular type of regression, linear regression (Figure 1.3) using least-squares, and nonlinear polynomial regression are both covered in Chapter 7.

Problems

1.1 Determine the type (linear or nonlinear) of algebraic equation assuming x, y, and z are unknown variables:

a) $x^2 + y^2 = 1.0$

b) $x + y = \sqrt{2}$

c) $y = 2 \cdot \sin(x)$

d) $x + y + z^2 = 0$

1.2 Determine the type (linear or nonlinear; ordinary or PDE) of differential equation assuming that z, x, and t are independent variables and g, D, and k are known parameters:

a) $\frac{d^2y}{dt^2} = -g$ (Newton's first law)

b) $\frac{\partial C_A}{\partial t} + v \cdot \frac{\partial C_A}{\partial z} + kC_A = \frac{\partial}{\partial z}\left(D\frac{\partial C_A}{\partial z}\right)$

c) $f'(x) = \sin(x) + 4$

1.3 If you want to determine the polynomial that interpolates 6 data points, what is the minimum order polynomial that is required? Write the polynomial with x as the independent variable and $a, b, c, \ldots$ as the unknown coefficients.

1.4 You are asked to use regression to determine the best linear polynomial fit for a given set of data. A colleague encourages you to determine the best fit by minimizing the sum of the distance between each point and the line instead of minimizing the sum of the *square* of the distance, which is the standard practice. The colleague claims that this will reduce the influence of a few outlying data points. Is the colleague correct?

1.5 You have been hired to produce an exact replacement part for a classic Porsche because the part is no longer available. Another engineer collects precise measurements of the location of a number of points on the surface of the part. You need to produce a new part with corresponding points at the same locations. Before machining the new part, you need to develop a continuous function that fits the measurement points because the continuous function will provide a representation of the surface connecting the points. Should you develop the continuous function using regression or interpolation between the precisely measured locations on the surface of the part? Why?

1.6 While studying a particular system, you collect some data on a measurable variable (y) versus an adjustable variable (x). Your next task is to use regression to approximately fit the data with a continuous mathematical function. Most engineers would start by trying to fit the data with a polynomial. You are not like most engineers because you wisely start by plotting the data. While examining the plot, you notice that the data has a pattern that is repeated as the adjustable variable is continuously changed. The measured variable increases and decreases regularly as the adjustable variable is increased. Should you fit this data with a polynomial? If so, what order polynomial? If not, what function(s) would you use instead?

Additional Resources

An understanding of how to solve differential equation problems is not required for understanding the material in this book. However, an ability to classify or recognize the type of equation that one is trying to solve is required. Most differential equation textbooks include a comprehensive set of definitions that enable the classification of mathematical equations. Some popular differential equation textbooks for engineers are:

- Differential Equations for Engineers and Scientists by Çengel and Palm [1]
- Advanced Engineering Mathematics by Zill and Cullen [2]
- Advanced Engineering Mathematics by Kreyszig [3]

and a helpful resource for data plotting and regression using Microsoft Excel is:

- Engineering with Excel by Larsen [4].

References

1 Çengel, Y. and Palm, W. III (2013) *Differential Equations for Engineers and Scientists*, McGraw-Hill, New York, NY, 1st edn.
2 Zill, D. and Cullen, M. (2006) *Advanced Engineering Mathematics*, Jones and Barlett, Sudbury, MA, 3rd edn.
3 Kreyszig, E. (2011) *Advanced Engineering Mathematics*, John Wiley and Sons, Inc., Hoboken, NJ, 10th edn.
4 Larsen, R.W. (2009) *Engineering with Excel*, Pearson Prentice Hall, Upper Saddle River, NJ, 3rd edn.

2

Programming with Python®

The objective of this chapter is to motivate the use of the Python programming language for solving problems in chemical and biological engineering and then to present a few basic principles associated with programming in Python. It is important to emphasize that the goal is not to cover all aspects of programming in Python because that would require an entire book (or potentially shelf of books) by itself. Instead, the goal is to present a few important principles and then slowly add additional Python programming knowledge throughout the remainder of the book.

2.1   Why Python?

When it comes to solving the many different mathematical problems that arise in engineering, many different software options exist for obtaining an exact or approximate solution. Some options, such as COMSOL or ANSYS, are very user-friendly and they hide most of the details of the calculations from the user. While these software packages represent an important resource for engineers, our goal here is, in fact, to learn and understand the calculations that are happening in the background of these commercial packages. We will not discuss these high-level software packages here simply because we want to focus on and understand the actual computational details.

Another set of software options for solving engineering problems are mathematical software packages such as MATLAB, Mathematica, or MathCAD. These packages give the user more control over the calculations, but they also require more specialized knowledge than the process simulation software described previously. These mathematical software packages are probably the most popular options for a college-level course on engineering calculations. They have one major disadvantage; however, they can be quite expensive, especially if the various supporting libraries and add-on packages are also required. It is true that many institutions have a site license for these software packages, but the license may require students to be on the school's network

Chemical and Biomedical Engineering Calculations Using Python®, First Edition. Jeffrey J. Heys.
© 2017 John Wiley & Sons, Inc. Published 2017 by John Wiley & Sons, Inc.
Companion Website: www.wiley.com/go/heys/engineeringcalculations_python

to use the software. It also means that the student is unlikely to have access to the software after they graduate.

The final option for the computational solution of engineering problems is to simply write your own computer code in a relatively low-level language such as FORTRAN or C++. Unfortunately, this option requires significant specialized knowledge – knowledge that is rarely retained beyond the course in which it is taught. Writing low-level computer code can also be a very frustrating experience when subtle errors in the code are difficult to identify due to obscure error messages. The result is that students spend most of their time looking for errors in the computer code instead of learning about computations and algorithm development.

There is not a perfect solution to the dilemma of selecting an optimal computer environment for learning computational techniques for solving engineering problems. However, the Python programming language has many advantages that make it the platform of choice here. These advantages include the following:

1) It is freely available and runs on most major computer platforms including Windows, MacOS, and Linux.
2) It has a tremendous number of additional libraries that are also free and add computational mathematics capabilities. For example, the Numpy library provides Python with capabilities that are similar to those of MATLAB.
3) It is an interpreted language (defined below) and is easier and faster for developing new algorithms than compiled languages.
4) Many libraries of previously compiled algorithms can be imported into Python, which allows for very fast and efficient computations.
5) It is worth repeating – it is free!

2.1.1 Compiled versus Interpreted Computer Languages

The first high-level programming languages that were developed, such as FORTRAN or C, were compiled languages. This meant that the programmer would type source code into the computer, this code was compiled into assembler code, and this was ultimately linked to produce a final executable file (Figure 2.1). The advantage of this approach is that the executable that was produced was relatively optimized and efficient for the platform on which it was built. Even today, most numerical software that requires significant computations, for example, meteorological software, is written in a compiled language. The disadvantage of this approach is that significant expertise and training are required to write computer programs in a compiled language, identifying errors in the source code is often a very difficult and time consuming process, and the resulting program can only be run on the platform or operating system for which it is compiled.

These disadvantages associated with compiled programming languages can largely be addressed through the use of interpreted languages. Common

Compiled languages:

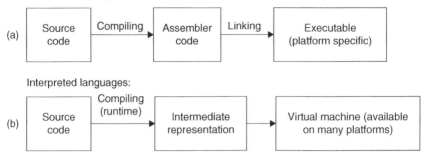

Interpreted languages:

Figure 2.1 The process of going from source code (i.e., a set of instructions) into a running computer program is different for compiled programming languages (a) versus interpreted programming languages (b).

interpreted programming languages include Java, Python, and JavaScript. Even MATLAB can be seen as an interpreted programming language. The source code for these languages is not compiled and linked to form a platform-specific executable but is, instead, compiled to an intermediate language (or bytecode) that is run on a "virtual machine." The virtual machine is a piece of software that interprets the bytecode and executes the instructions contained in the original source code. One obvious advantage of this approach is that the source code can be run on any computer that has the required virtual machine. Since Python and many associated libraries are available for all the major operating systems, you can execute Python source code almost anywhere. Interpreted languages also tend to be easier to program with because the syntax is more forgiving and the error messages are more informative (although you will still see cryptic error messages and frustrating syntax requirements in all computer languages). The disadvantage of interpreted languages is that they tend to execute instructions more slowly than compiled languages – often by a factor of 10 or more. If we need to multiply 10^{14} numbers by π, a factor of 10 can mean the difference between a 1 h computation and a 10 h computation. Interpreted languages are getting faster all the time, however, and they are starting to close the gap between compiled and interpreted languages. One common strategy is "just-in-time" (JIT) compilation. The basic idea here is that the virtual machine can actually compile important and frequently run source code all the way to a platform-specific executable (just like a compiled language). Of course, this "on-the-fly" compiling slows down the execution of the rest of the computer program, but, if a particular set of instructions is executed frequently, it may be more than worth the cost of JIT compilation.

2.1.2 A Note on Python Versions

In 2008, a new version of Python, Python 3.0, was released. This new version contained a few significant changes from the previous Python 2.x series.

In particular, programs written for the Python 3.x series would normally not run on the Python 2.6 and earlier series of virtual machines, and existing programs written for Python 2.x virtual machines would not run on Python 3.x series virtual machines. Probably, the biggest change impacting the Python codes in this book has been to the "print" function notation. In Python 2.6 and earlier versions, the format was

```
print "Hello World"
```

and for Python 3.0 and later versions, the format was

```
print("Hello World")
```

Those parentheses may look like a small change, but the new format is not compatible with Python 2.6 and earlier versions, and the old format is not compatible with Python 3.0 and later versions. Interestingly, Python 2.7 supports both versions. It should also be noted that there were other, more significant changes beyond the print function when the change was made to Python 3, but those changes rarely impact the codes and types of algorithms written in this book.

As of 2015, most numerical python libraries are available for Python 2.7 or Python 3.x virtual machines. The examples in this textbook were written for a Python 3.x series virtual machine but have also been tested on a Python 2.7 series virtual machine. It is inevitable that all Python computations will eventually transition to Python 3.x or later virtual machines. In the meantime, it is important to recognize the version of Python that you are using and select the appropriate virtual machine for the code that is being executed.

If you ever need to determine the version of Python that you are currently using, you can type the following two Python commands:

```
import sys
print(sys.version)
```

The system that I am currently using prints out

```
Python 3.4.0 (default, Jun 19 2016, 14:20:21)
[GCC 4.8.2] on linux
```

2.2 Getting Python

The process of learning numerical methods for engineering requires writing and executing computer programs. This book advocates the use of Python for writing and executing these computer programs so it is highly recommended that the reader have access to at least Python 2.7 (although Python 3.4 or later is recommended) plus the following libraries:

- Numpy (www.numpy.org) – array operation library
- Scipy (www.scipy.org) – scientific algorithm library that uses numpy

- Matplotlib (www.matplotlib.org) – provides the pyplot and pylab plotting libraries
- SymPy (www.sympy.org) – symbolic mathematics library (optional, used primarily in Chapter 5)
- Pandas (http://pandas.pydata.org/) – easy to use data structures and data analysis tools including data import (optional, used primarily in Chapter 9)

It is also recommended that an integrated development environment (IDE) be used to facilitate the writing of Python Source code. One particularly good IDE is called Spyder (https://pythonhosted.org/spyder/). Figure 2.2 shows the basic layout of the Spyder IDE interface. The input window on the left side of the Spyder program window shows the Python source code that is currently being edited. The code in the source window can be executed or run by selecting "Run" from the "Run" menu or simply pressing F5 on most platforms. The upper right-hand screen usually shows documentation when it is available for different functions included with Python or imported libraries. The lower high-hand screen shows a Python console or Python prompt, ">>>". Basically, the Python prompt is an actively running Python virtual machine and different Python commands can be tested at the prompt.

When writing a new Python program, it is often helpful to "try out" a command or line of code at the Python prompt to observe the result. Having an active virtual machine for testing ideas helps to make Python an efficient language for writing new programs.

Figure 2.2 A screenshot of the Spyder IDE for Python programming including source code window on the left size, documentation window on the upper right side and Python console for rapid testing and executing the source code in the lower right side.

2.2.1 Installation of Python

For computers running Windows, three good options for installing Python include the following:

- Anaconda Scientific Python (store.continuum.io/cshop/anaconda/)
- pythonxy (code.google.com/p/pythonxy/)
- winpython (winpython.sourceforge.net)

All of these packages include Python plus all the required libraries such as numpy and scipy plus they include the Spyder IDE. As of 2015, only Anaconda Scientific Python supported Python3. Presumably, the other options will eventually support Python3, but care should be taken when installing Python to select the desired Python version – 3.x or 2.x.

For computers running MacOS, it is easiest to install the Anaconda Scientific Python package (https://store.continuum.io/cshop/anaconda/).

For computers running a Debian-based version of Linux, the following command will install all required libraries:

```
sudo apt-get install python3-numpy python3-scipy
sudo apt-get install python3-matplotlib ipython3
```

The FEniCS program, which is used in Chapter 14, is only available for Debian-based versions of Linux (i.e., Ubuntu or Mint Linux) or Docker and can be installed on Debian systems using

```
sudo apt-get install fenics
```

It should be noted that as of 2016, FEniCS requires Python 2.7, but it is expected to move to Python 3.x in the near future.

First Python Commands!

Open a Python console or open the Spyder IDE and move the cursor down to the lower right corner. At the Python prompt, ">>>" type,

```
>>> print('Hello World')
```

and the console should print "Hello World" back to the screen. Note that you do not type the ">>>" prompt as it should automatically appear within the open console. If you are using an iPython console, the prompt will look like: In [#]: where # is an integer, and the program above would be:

```
In [1]: print('Hello World')
```

In general, the regular Python prompt: ">>>" and the iPython prompt: In [#]: give the some behavior although the iPython prompt supports more commands.

(Continued)

First Python Commands! (Continued)

For a slightly fancier version of the example above, set the variable "a" equal to the string "hello," set the variable "b" equal to the string "world" (note the space and the beginning), and then ask Python to "print(a+b)". The exact sequence of instructions should give

```
>>> a='hello'
>>> b=' world'
>>> print(a+b)
hello world
```

Congratulations if you just executed your first Python program!

2.2.2   Alternative to Installation: SageMathCloud

If a Windows, Mac, or Linux computer is not available for installing Python and the important scientific libraries, most of the material covered in this book, including examples, exercises, and problems, can be completed using SageMathCloud. SageMathCloud is a web-based computing platform for computational mathematics, and it is part of the Sage project. Basically, the SageMathCloud project has installed a large number of software packages, including Python and the libraries used in this book, on computers connected to the internet, and then they provided a web-based interface to this software. The result is that users can visit the SageMathCloud website, cloud.sagemath.com, create an account, and then start writing scientific software in Python (or other languages including Julia, R, and Octave) within the webpage. The website can be used from almost any web-browser, including smartphones, tablets, and chromebook (ChromeOS) computers.

The web-based interface is based on the Jupyter project. Users start by creating a new project, then creating a new Jupyter notebook. An example of a Jupyter notebook is shown in Figure 2.3. Jupyter notebooks consist of cells that contain one or more lines of Python code. The code within a cell is executed by pressing "Shift-Enter", and once the code is executed, the results and other output are displayed below the cell and stored in memory (more precisely, all python objects are retained and can be used when executing other cells). Overall, the style of Jupyter notebooks is similar to Mathematica Notebooks for individuals that are familiar with that software.

With all the benefits associated with SageMathCloud, including the ability to write and execute scientific Python code from anywhere, including a tablet, one might ask why Python should ever be installed on a computer? Why not always use SageMathCloud? There are a few reasons. First, SageMathCloud requires an internet connection and if that connection is lost, work can be lost. Second, at times of heavy use, the internet-connected computers that are

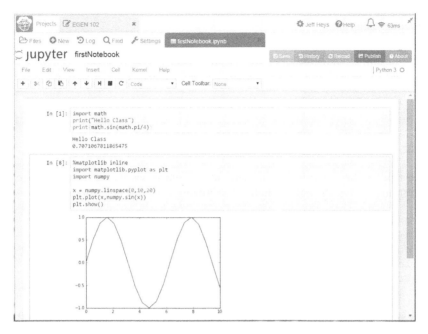

Figure 2.3 A screenshot of the Jupyter notebook on SageMathCloud. Two different cells are populated with Python code, and the cells are executed using "Shift-Enter". The results of code execution are shown below each cell.

actually running the Python code for SageMathCloud can become slow – up to a factor of 10 slower than a modern laptop running Python. Finally, and this is the biggest reason, it can be difficult or impossible to use a Python library that is not already on SageMathCloud. For example, the finite element library FEniCS that is covered in the last chapter of this book is not available on SageMath-Cloud. If you are using your own computer, you can install it yourself, but, if you are using a web-based interface into some other computer, that may not be possible.

2.3 Python Variables and Operators

Programming frequently requires us to assign a variable to a specific piece of data (or something more complex). For example, typing:

```
a = "hello"
```

into the console or a Python script file results in the variable "a" being *assigned* to character string "hello". The word *assigned* is emphasized here because it better reflects the role being played by the equal sign. Whenever Python code contains "=", the object on the right is being assigned to the variable on the left.

Variable Assignment

In Python (and most other programming languages) we should see:

  a = "hello"

as

  a ← "hello"

The role of the assignment operator may seem obvious, but many novice programmers have struggled when the following code did not work:

```
>>> a=4
>>> a=b
Traceback (most recent call last):
  File "<stdin>", line 1, in <module>
NameError: name 'b' is not defined
>>>
```

The novice programmer may believe that the second line ("a=b") will result in "b" being set to 4 since "a" was previously set to 4. This will not happen, and, instead, we get an error because the result of executing the code is that "a" is assigned to something that is not defined (the variable "b" has not been assigned). Notice that the end of the Python error message is telling us the problem.

Simultaneous Assignment

Python also allows the simultaneous assignment of multiple variable to the same value. For example,

  a = b = 0

is the same as

  a ← b ← 0

and both a and b are assigned the value of 0.

Python uses *strong-typing* for variables, which means that every variable is a specific type, for example, an integer, floating point number, and character. There is a built-in function in Python called *type()*, which will return the type for a given variable. In some cases, it is possible to convert from one variable type into another variable type as is illustrated in the example below where a string ("str") variable is converted into an integer ("int").

```
>>> a='5'
>>> print(type(a))
<class 'str'>
>>> c=int(a)
>>> print(type(c))
<class 'int'>
>>> print(type(3.1415))
<class 'float'>
```

The Python code above was entered into a Python console. If you try this example for yourself, do not type the ">>>" prompt, it should be part of the console. The output from the print statement might vary slightly depending on the operating system, Python version, and type of Python console you are using, but the results should contain "str", "int", and "float". You can also enter the above code into a text file or script. For example, you could enter the code:

```
a = 5
print(type(a))
c = int(a)
print(type(c))
print(type(3.1415))
```

into the left half of the Spyder IDE window (or any other program that can edit text files), save the file, and then run it through the Python virtual machine. The output should be the same as the previous series of commands at the console: "str", "int", and "float". Examples of both types of code entry: at the console or in a script, are shown in Figure 2.4.

2.3.1 Updating Variables

When writing a computer program, it is often necessary to update the value of a variable. For example, we may want to count something by initially setting a variable to zero, and then adding one to the value of the variable as we count, for example, the number of words in a paragraph. In order to add one to a variable, we could type:

```
>>> w = w + 1
```

where w is the variable that holds the number of words as we count. The line of Python code above is clearly completely invalid from a mathematical standpoint. It "looks" like an equation where w is equal to itself plus one – a mathematically unsolvable equation (unless zero equals one!). However, it is NOT a mathematical equation, it is NOT a statement of equality, it IS an assignment: w is assigned a value of the previous value of w plus one. This line should be viewed as:

$$w_{new} \leftarrow w_{old} + 1$$

This type of variable update – where a variable's value is updated by adding something to it – is so common in programming, that a special notation is

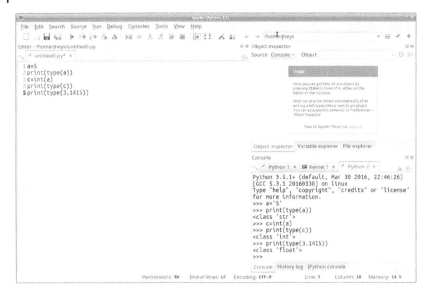

Figure 2.4 A screenshot of the Spyder IDE showing the two different methods for entering the example Python code: at the console in the lower right corner or into a script (i.e., text file) in the left half. The script on the left is run using the green, triangular "play" button along the upper part of the window.

available in Python for it. To add the value of v to the value of w and store the result in w, we may write

```
>>> W += V
```

and this is identical to

```
>>> W = W + V
```

The new value for w will equal the old value for w plus the value of v in either case.

Mathematical Operations

Now that we know how to assign variables, we are in position to explore operators like "+" and "*", which allow us to effectively use Python like a calculator. The following example illustrates some features of operators.

The following can be typed into the Python console:

```
>>> a=4
>>> b=2
>>> print(a+b)
6
>>> print(a-b)
2
```

multiplication and division:

```
>>> print(a*b)
8
>>> print(a/b)
2.0
```

exponent, floor division, and remainder:

```
>>> print(a**2)
16
>>> print(b%a)
2
>>> print(a%b)
0
>>> print(9//b)
4
```

2.3.2 Containers

It is often useful in programming to collect multiple objects together into a single container and assign them to a variable. Python includes a number of different types of containers including tuples, lists, and dictionaries. The focus here is on numerical computations and the most useful type of container for these algorithms is a list container. In Python, a list has one or more objects (usually numbers for numerical computations) separated by commas and surrounded by square brackets. Lists may be heterogeneous – containing different objects types, but in practice, most lists only contain one type of variable. Lists should remind us of vectors. The construction of lists is illustrated below.

```
>>> myList = [1, 2, 3]
>>> print(myList)
[1, 2, 3]
>>> secondList = ['a', 'b', 'c', 1, 2, 3]
>>> print(secondList)
['a', 'b', 'c', 1, 2, 3]
```

The two lists created above were stored in two different variables, "myList" and "secondList". It is common to store lists in variables. We often wish to modify lists that have already been created. Some examples of list modification are shown below.

```
>>> vec1 = [2, 3, 5]
>>> vec2 = [24, 2, 10]
>>> vec1.append([8,2])
>>> print(vec1)
[2, 3, 5, [8, 2]]
```

```
>>> vec1[3] = 87
>>> print(vec1)
[2, 3, 5, 87]
>>> vec2.extend([9,3])
>>> print(vec2)
[24, 2, 10, 9, 3]
>>> print(vec1+vec2)
[2, 3, 5, 87, 24, 2, 10]
>>> print(vec1+"what?")
Traceback (most recent call last):
  File "<stdin>", line 1, in <module>
TypeError: can only concatenate list (not "str") to
list
```

Two lists are defined at the console and assigned to the variables "vec1" and "vec2." To access an item in a list, use square brackets after the name of the variable, for example, vec1[3] accesses the fourth item in the list. It is important to emphasize that in Python (and many other modern programming languages, the first item is a list has the index zero. To access the first item in vec1, use vec1[0]. Counting from zero can be awkward at first, but most experienced programmers appreciate the subtle advantages that will hopefully become apparent later.

After defining the lists, a nested list is then appended onto the end of "vec1" and then the nested list is replaced with 87 (programmers like to say that Python lists are mutable and can be changed). Lists can also be extended as "vec2" is extended, or lists of the same type can be concatenated as "vec1" and "vec2" are combined together. A list cannot be concatenated with a string or almost anything other than another list as evidenced by the error message at the end.

Tuples are another type of container in Python that are, in practice, very similar to lists, they are an ordered collection of objects that may be heterogeneous. The main difference is that tuples may not be changed. They are immutable. A simple demonstration of tuples is shown below:

```
>>> mytuple = ('a', 1, 'c')
>>> print(mytuple[2])
c
>>> (x,y,z) = mytuple
>>> print(x)
a
>>> (j, h) = mytuple
Traceback (most recent call last):
    (j, h) = mytuple
ValueError: too many values to unpack (expected 2)
```

Notice how the individual objects within the tuple are still accessed with square brackets. In addition, observe that the individual objects within the tuple can be assigned to a set of individual variables, in this case named "x", "y", and "z". This is only practical for tuples with small numbers of objects, but it can be useful as we will see in the next chapter. Of course, if you try to assign three individual objects to only two variables, the result is an error.

One final type of Python container that is only briefly mentioned here is the dictionary container. With lists and tuples, the objects in the container were in a specific order, and we could access the objects using an integer corresponding to the location of the desired object within that order. The first and third objects in a list or a tuple are accessed using "name[0]" and "name[2]," respectively, if the list or tuple is stored in variable "name." Dictionaries are containers that contain a potentially heterogeneous collection of objects, but dictionaries are *not* ordered. Instead, a *key* is assigned to access the object. The key can be a string or an integer or other descriptor, and when building the dictionary, the key is followed by a colon (:) and then the object that the key is referencing. To access the object referenced by the key, square brackets containing the key are used. The example below builds two dictionaries, one uses strings for the keys and the other uses integers for the keys:

```
>>> myDict = {'class':'102', 'instructor':'Heys'}
>>> print(myDict['instructor'])
Heys
>>> secondDict = {0:'zero', 2:4}
>>> print(secondDict[0])
zero
```

Notice that dictionaries are constructed using curly brackets, "{" and "}," and every entry is a "key:object" pair. Using integers as keys in the second example causes the dictionary to look a little like an ordered list or tuple but that is not the cases. If one tried to access `secondDict[1]`, for example, an error would result. Dictionaries are much less common in engineering computations and are not used in this book for the algorithms presented.

2.4 External Libraries

Imagine that we want to calculate sin(1.2). If we try typing that into the Python console or a simple piece of source code, this is what we are likely to see:

```
>>> sin(1.2)
Traceback (most recent call last):
  File "<stdin>", line 1, in <module>
NameError: name 'sin' is not defined
```

We get an error message because the sin() function is not a built-in function in Python. In order to use the sin() function, we need to import the "math" library into Python. This can be accomplished two different ways and both are shown below.

To import a library, use the **import** command

```
>>> import math
>>> math.sin(1.2)
0.9320390859672263
```

This is the preferred approach. The first command imports the entire math library, and any functions, methods, or data contained within that library can be accessed by typing: math.name or math.name() where name() is the name of a function in the library. A complete list of functions and values within the math library can be found at docs.python.org/3/library/math.html.

The other approach to importing a library uses the **from** command

```
>>> from math import *
>>> sin(1.2)
0.9320390859672263
```

This approach loads the entire math library into the global Python namespace, which can be thought of as a list of reserved words that are already defined. For example, **import** is a reserved word that is part of the global Python namespace and we should never use **import** for any other purpose. For example, do not try to use "import" as a variable name. Whenever we load a library into this same global namespace, we greatly increase the number of global terms and invite the possibility of conflict. For example, if we tried to load two libraries that both contain the sin() function (both the math library and the Numpy library, which we use frequently both contain a sin() function), Python would give us an error. There are times when it is easier to use the **from** name **import** * option for loading libraries, but it is usually better to use **import** name.

If you tried to run >>> sin(1.2) at the console and were successful, this was a result of using an IDE that was smart enough to load the math library for you.

There are hundreds of libraries that have been written by others that increase the power of Python and save us from having to rewrite code that has already been written by others. In this book, we will use the *math, numpy, scipy,* and *matplotlib* libraries extensively. Two libraries that are not covered in great depth in this book but maybe helpful in further developing engineering algorithms are the "sys" library and the "timing" library. The sys library provides information and functions for connecting with the computer's operating system. For example, **print**(sys.platform) will print out a string that identifies the underlying operating system. The sys.float_info data structure has information about floating point precision and computational roundoff for the computer currently being used. The "time" library can be useful for measuring the time required to run various parts of an algorithm.

The function `time.clock()` returns the current processor time (in seconds), and by storing the time in different variables and then calculating the difference between those variables, it is possible to determine the processor time required to execute a series of commands. For example, the following code:

```
t0 = time.clock()
...Python code to be timed...
print(time.clock() - t0, "seconds process time")
```

will print out the time required to run the Python code in between the two calls to `time.clock()`. This information can be especially helpful in determining how an algorithm's computational cost scales with the change in a parameter or in determining the slower sections of an algorithm's execution.

2.4.1 Finding Documentation

One of the challenges associated with external libraries (and one of the general challenges associated with programming) is finding documentation that describes how to use the various functions, methods, and data associated with the library or programming language. For example, where can we find data about the sin() function that is part of the math library? For most programmers, the easiest answer is just to perform an internet search. To search for documentation about the sine function, a search for "python math sin" would probably yield abundant documentation.

Python has the benefit of supporting *docstrings*, which often provide additional help for the use of functions and methods in external libraries. The use of the help() function to access *docstrings* is illustrated below.

```
>>> import math
>>> help(math.sin)
Help on built-in function sin in module math:

sin(...)
    sin(x)

    Return the sine of x (measured in radians).
```

Another challenge when using external libraries is that we sometime wish to have a list of available functions and data that are part of the library. A directory or list of items in a library can be accessed through the dir() function. For example, a list of functions and data available in the math library can be accessed as shown below.

```
>>> import math
>>> dir(math)
['__doc__', '__name__', '__package__', 'acos', 'acosh',
'asin', 'asinh', 'atan', 'atan2', 'atanh', 'ceil',
```

```
'copysign', 'cos', 'cosh', 'degrees', 'e', 'erf',
'erfc', 'exp', 'expm1', 'fabs', 'factorial', 'floor',
'fmod', 'frexp', 'fsum', 'gamma', 'hypot', 'isinf',
'isnan', 'ldexp', 'lgamma', 'log', 'log10', 'log1p',
'modf', 'pi', 'pow', 'radians', 'sin', 'sinh', 'sqrt',
'tan', 'tanh', 'trunc']
```

Note that some of the items in the math library are functions, like "sqrt" or "exp," and some of the items are constants, like "pi".

Problems

2.1 Using Python, calculate 5/16, 5.0/16, and 0.5^2. Include all Python commands and results (using a cut-and-paste approach is recommended). Are the results correct. For Python versions before Python 3.x, integer division returned an integer result. As a result, 5/16 would return 0 instead of the floating point answer (0.3125). Whenever you use division in a Python algorithm, it can be important to make sure that one of the numbers is a floating point number or multiple the numerator by 1.0 to convert it into a floating point number. Alternative, in Python 2.6 and 2.7 codes, you will often see the line: **from**__future__ **import** division to get the same behavior as Python 3.x where the result is always returned as a floating point number.

2.2 When using containers in Python, variable assignment can give some interesting behavior. Begin by creating a list of integers from 1 to 5, and store the list in variable "a". Next, assign the variable b to be equal to a. Finally, change the value of the second entry in the list of integers to the number 20. The code below summarizes the steps necessary.

```
>>> a=[1,2,3,4,5]
>>> print(a)
>>> b=a
>>> b[1]=20
>>> print(b)
>>> print(a)
```

Summarize the output of the code above and describe the behavior of assigning variable "b" equal to "a".
For the second part of the problem, repeat all the steps in first part, except assign "b" to all the values within "a" using

```
>>> b=a[:]
```

Again, summarize the output of the code and describe the assignment behavior.

2.3 Another type of Python container that is less common in engineering computations is the tuple. Tuples use rounded brackets: "(" and ")" instead of square brackets, "[" and "]" like lists. The main difference between tuples and lists is that tuples are immutable, that is, they cannot be changed. Try to repeat the first few commands from the previous problem using a tuple instead of a list:

```
>>> a=(1,2,3,4,5)
>>> print(a)
>>> print(a[1])
>>> b=a
>>> print(b)
>>> b[1]=20
```

What did you observe? Can you think of a way to get a new tuple from an previous tuple with a number that is changed?

2.4 Assign a variable to a string that is your name. For example, I would perform the following assignment:

```
a='jeff'
```

Write the Python command that will print the second letter (e.g., for me, it should print the letter "e"). Now, try to replace the second letter in your name with the letter "z". Describe and explain what you observe.

2.5 While writing a Python program, you decide to assign a new variable "class" to the value of 1,2,3, or 4 depending on the current class year for a group of students. In Python, try to assign the variable "class" to a value of 1. Describe what you observe. Next try to assign the variable "pass" to a value of True. Describe what you observe.

To better understand what is happening, type "help()" at the Python prompt (i.e., the >>> prompt). This should bring up the help prompt (i.e., "help>"). At this prompt, type "keywords" to get a list of reserved keywords that cannot be used as variables. Print this list of Python keywords.

Additional Resources

Books on general programming in Python:

- Learning Python by Lutz [1]
- Python in a Nutshell by Martelli [2]
- Think Python by Downey [3].

References

1 Lutz, M. (2013) *Learning Python*, O'Reilly Media, Inc., Sebastopol, CA, 4th edn.

2 Martelli, A. (2009) *Python in a Nutshell*, O'Reilly Media, Inc., Sebastopol, CA, 2nd edn.

3 Downey, A. (2012) *Think Python*, O'Reilly Media, Inc., Sebastopol, CA, 1st edn.

3

Programming Basics

The objective of this chapter is to continue learning the basics of programming in Python. In the previous chapter, we learned how to assign variables to different values, including integers, floating point numbers, and strings. We also learned how to import additional functionality from external libraries like the math library. This chapter covers additional, standard topics in programming like logic, looping, conditionals, and developing our own functions.

3.1   Comparators and Conditionals

As described previously, the equal sign does not actually compare two objects to see if they are equal. If we wish to compare two objects for equality, we need to use == as illustrated:

```
>>> a=4
>>> b=4
>>> a==b
True
>>> a<b
False
>>> a != 2
True
```

Beyond the equality comparator, the less than, "<", greater than, ">", less than or equal to, "<=", greater than or equal to, ">=", and NOT equal, "!=", comparators are frequently helpful. In all cases, the comparator should return a Boolean, `True` or `False`.

While the focus in this textbook is on numerical programming, it can be interesting to try out some of the same principles on strings of characters. Consider the following:

```
>>> a="hello"
>>> b="world"
```

Chemical and Biomedical Engineering Calculations Using Python®, First Edition. Jeffrey J. Heys.
© 2017 John Wiley & Sons, Inc. Published 2017 by John Wiley & Sons, Inc.
Companion Website: www.wiley.com/go/heys/engineeringcalculations_python

```
>>> a==b
False
>>> a<b
True
>>> b<a
False
```

Here, the comparator compares two strings to determine which is first alphabetically.

Comparators can be extremely helpful in constructing conditional statements. For example, if we want a block of code to only execute when a certain condition is true, we can use an **if** statement:

```
>>> a=4
>>> if a<5:
...     print("smaller")
...
smaller
```

or, in the form of a script:

```
a=4
if a<3:
    print("smaller")
else:
    print("larger")
```

where the code will, of course, print "larger" upon execution. From these two examples, we can make an *INCREDIBLY IMPORTANT OBSERVATION* (note that I wish that I could make the next sentence flash). In Python, blocks of code are designated using indentation. Every **if** statement has a condition followed by a colon. If the conditional is True the following block of text is executed, and the scope or length of the block is determined by the fact that all lines of code within the block MUST be indented EXACTLY the SAME amount. If the first line in the block is indented four spaces (and four spaces is the standard Python style), then every line must be indented four space. If you mess up and indent one line with only three spaces or a tab, error messages and chaos will follow. Most people find indentation at the Python prompt (>>>) awkward and difficult. Any algorithm that requires indenting a block of code is probably sufficient long that it should be developed using a script and not simply entered at the Python prompt. The same consistent indentation requirement for one block of code following a conditional statement also extends to nested conditional statements, as shown in the example below.

```
a=input('Enter an integer (0-10): ')
# convert the input to an integer if possible
a = int(a)
if a<5:
    print("a is less than 5")
```

```
    print("adding 1")
    a += 1 # this is identical to a=a+1
    if a<4:
        print("a is still less than 4")
        print("adding 1 more")
        a += 1
print("a = ", a)
```

The goal of this simple code is to increase the value entered if it is "small". If the value is less than 5, then 1 is first added to the original input value. If the value is still less than 4, a second 1 is added. Upon execution with an input of "4", this code generates

```
a is less than 5
adding 1
a =  5
```

and an input of "2" generates

```
Enter an integer: 2
a is less than 5
adding 1
a is still less than 4
adding 1 more
a =  4
```

In the previous example, we can see an example of a "comment" being include with the Python script. The comment, "`# this is identical to a=a+1`", is included in the program to make it more readable and easier to understand. The use of frequent and descriptive comments is highly recommended. A good rule of thumb is that one comment should be included for every two lines of regular Python code. Another good rule of thumb is to use roughly an order of magnitude more comments in your own code compared to what you will find in this book! In Python, a comment is initiated by the "#" character and all following characters are not interpreted or executed by the Python virtual machine – it is as if they do not exist. Optionally, multiline comments can be initiated use three consecutive double quotes and ended using three consecutive double quotes. One final note, the comment character, "#", can also be helpful for temporarily removing a line of code from execution.

One new Python function used in the previous example is the `input()` function for getting input from the keyboard. The input from the keyboard is stored in the variable a in the example. It is *always* a good programming practice to check the validity of input each and every time. In the above example, it would be good to check that a is between some minimum and maximum integers before using the variable any further.

Whenever a comparator is used, Python returns a Boolean (i.e., `True` or `False` in Python). The Boolean can be stored in a variable as is illustrated in a simple example:

```
j = 3 < 4
print(j)
if j:
    print('true')
```

and we note that the result can be used in a conditional **if** statement. The output of this script should be

```
True
true
```

because the **if** j: comparator is identical to **if** 3<4: in this example.

3.2 Iterators and Loops

When creating computer algorithms, we often need to repeat a series of commands or instruction a number of times. For example, we might want to compare the value of a variable to each individual in a list of values. In numerical algorithms, we might need to take the sin() of each number in a list of numbers. For all these situations (and many more), we need to use iterators. The most common types of iterators (or loops) that are used in this book are the **for** loop and the **while** loop.

"for" Loops

An extremely simple **for** loop is

```
for n in [2,3,4]:
    print(n)
```

and the output from running this loop is

```
2
3
4
```

   Notice that the structure of the for command is

```
for variable in list:
```

where the variable takes the value of each item in the list in order. Also notice that the **for** command ends with a colon (:) and the block of code to be executed each iteration through the loop is indented.

The simple example above illustrates the basic elements of a **for** loop, but what if you wanted a loop that repeated 50 times or 50 million times? Would we need to type out a list of 50 million numbers? The answer is, of course, "No!"

Python conveniently provides the **range()** function for iteratively generating lists of numbers of any desired length. If a single number is passed into the **range(n)** function, it will iteratively generate a list of integers from 0 to $n − 1$ or 1 less than the number that is passed into the function. Since the list starts at 0, the command **range(5)** will generate a list of five integers: 0, 1, 2, 3, 4.

Let us illustrate this by constructing a **for** loop that executes 5 times and calculates the sin() of each integer between 0 and 4.

```
import math

for i in range(5):
    # i is 0, 1, 2, 3, 4
    j = math.sin(i)   # input in radians
    print(i, j)
print("finished")
```

Upon execution, the output from this code should be

```
0 0.0
1 0.8414709848078965
2 0.9092974268256817
3 0.1411200080598672
4 -0.7568024953079282
finished
```

It is also possible to have a **for** loop that iterates through a list of something other than integers. Consider the following example:

```
for color in ["red", "yellow", "green"]:
    signal = color + "light"
    print("The signal shows:", signal)
```

When this script is run, the output should be

```
The signal shows: red light
The signal shows: yellow light
The signal shows: green light
```

Similarly, you can iterate through a list of lists, and even store the results in two different temporary variables, as illustrated in this example:

```
for m,n in [[1,2],[3,4],[5,6]]:
    # notice change of order for m,n below
    print("n = ",n," and m = ",m)
```

and the output from this loop is

```
n =    2   and m = 1
n =    4   and m = 3
n =    6   and m = 5
```

Python makes it easy to iterate through any list and perform various operations each iteration.

It is often necessary to nest a conditional within an iterator. The outer loop consists of some list that we are iterating through, and the inner loop consists of a condition that is executed whenever the condition is met. A simple example of this would be to iterate through a list of integers and then print out the value of integers if they have 7 as a factor (i.e., they are evenly divisible by 7). The simple Python code below performs this task.

```python
for i in range(100):
    if i%7 == 0:
        print(i)
```

The output is a list of integers, starting with zero and then counting by 7 up to 98. We could also try the slightly more complicated task of determining all the integers between 0 and 150 that have 5 or 7 as factors. This change requires the use of a logical **or** in the **if** statement. We also wish to count the total number of integers that meet these criteria.

```python
counter = 0
for i in range(150):
    if (i%7 == 0) or (i%5 == 0):
        counter += 1 # increment the counter
        print(i)
print('Total: , counter)
```

The total number of integers between 0 and 150 that have 5 or 7 as factors is 47, which is printed out at the last line in the script. A very common mistake made by novice programmers is to indent that last line. Whenever you are iterating and accumulating or totaling something as the iterations occur, it is critical that no action be taken until the iteration has completed. If that last line is indented, Python does not know that the programmer really wanted to wait till the end to print the result to the screen.

Next is an example of a **while** loop that continues until a specific condition is met. Two warnings regarding **while** loops: (1) make sure that all variables in the stopping condition are initialized to starting values before the while statement and (2) make sure that the loop cannot repeat for an infinite number of times.

```python
import math

x=0.0
while x < 10.0:
    y = 2.0
    x += y
    print(x,y)
```

Upon execution, the output from this code should be

```
(2.0, 2.0)
(4.0, 2.0)
(6.0, 2.0)
(8.0, 2.0)
(10.0, 2.0)
```

While loops are often used in situations where an unknown number of iterations is required. It is critical in these situations, however, to have some type of stopping condition to prevent the loop from iterating or executing forever. Imagine that we wanted to generate random numbers between 0.0 and 1.0 and then count the total number of random values generated before generating a random value that exceeds some threshold. These types of calculations are common when developing mathematical models of stochastic or random processes like chemical kinetics. The Python code below allows us to count the number of random values generated before exceeding a set threshold.

```python
import random

maxIter = 10
currentIter = 1
val = random.random() # returns a random number 0-1
print(val)

while (currentIter < maxIter) and (val < 0.75):
    currentIter += 1 # increment the counter by 1
    val = random.random()   # new random number
    print(val)

if (currentIter < maxIter):
    print('Iterations required for a random number')
    print('greater than 0.75? ', currentIter)
else:
    print(maxIter, ' iterations reached before a ')
    print('value greater than 0.75!')
```

The variable currentIter is a counter for the total number of random numbers generated, and val is the value of the most recent random number. The key line in this code is the **while**-loop line. The block of code (3 lines total) below the **while** is repeated as long as two different conditions are both true: (1) the iteration counter is less than the variable maxIter and (2) the most recent random number is less than 0.75. Since one out of every four random numbers should be larger than 0.75, it is rare that the maximum

number of allowed iterations is reached, but it can and does happen. A typical
output from the code is shown below:

```
0.3469218564154297
0.052444292456515496
0.31212308088974927
0.031893177151028684
0.63273646126271
0.8911860542672484
Iterations required for a random number (0-1)
greater than 0.75?   6
```

The output is different every time the code is executed because different
random numbers are generated each time, although it is possible to modify
the code so that the same set of "random" numbers is generated each run. In
addition, as we will see in the next chapter, it is probably more efficient to
generate a list of 10 random numbers in a single call to a random number
generator and then loop through the list until we reach one larger than the
threshold (0.75 for the example above).

It is possible to use iterators to generate more complex lists. For example, if
we want a list of perfect squares, then we can put an iterator inside the square
brackets normally used for defining a list. For example,

```
x2 = [x*x for x in range(6)]
print(x2)
```

will generate the list [0, 1, 4, 9, 16, 25]. These types of operations
are called *list comprehensions*. They can also be useful for generating sublists
using comparators. For example, the list of factors of 3 from the full list of integers:

```
x2 = [x for x in range(20) if x % 3 == 0]
```

will generate the list [0, 3, 6, 9, 12, 15, 18]. Finally, the Python
keyword **in** may be used to check for membership in a list. The previous script
can be used to check if specific values are in the resulting list.

```
x2 = [x for x in range(20) if x % 3 == 0]
print(9 in x2)
print(10 in x2)
```

The script will generate True and then False.

The range() Function

The `range` function in Python supports the following arguments:
`range(start, stop, step)`. If only one input is given, it is treated as the
stop value and the function returns the integers from 0 to `stop-1`. If two values

are given, they are treated as start and stop values, and three values are treated as start, stop, and step size. Using this information, try to construct a loop that prints out the odd integers from 7 to 17, including 17.

We need to set `start` to 7, `step` should be set to 2 so that we only have odd integers, but how do we get the loop to include 17, but exclude 19?

The following will not work:

```
for i in range(7, 17, 2):
    print(i)
print("finished")
```

This code will stop at 15. Instead, because **range** does not include the value of `stop` in the sequence, we need to set `stop` to 18 or 19. You may also want to try setting the value of `start` to 7.5 or some other non-integer. The result will not be good because **range** requires all input values to be integers.

3.2.1 Indentation Style

Before leaving the topics of loops and conditionals, it is worth revisiting an important issue that was introduced in this section – indentation. In Python, blocks of code are designated by a common level of indentation. If we have multiple lines of code that should be executed as part of a **for** loop or as part of an **if** statement, then those lines must have a uniform level of indentation. While Python is agnostic about the type – spaces or tabs – and the quantity used for indentation, I recommend that novice programmers follow the style recommended in the Style Guide for Python Code, also known as Python Enhancement Proposal #8 or PEP 8 [1]. Specific style guidelines include the following:

- use spaces instead of tabs for indentation,
- use four spaces for each level of indentation,
- try to keep lines to 79 characters or less,
- do not put spaces around list or array indices, that is, use a [5] instead of a [5], and
- put one, and only one, space before and after the equal sign ("=") in variable assignments.

Additional Python style guidelines, including best Python programming practices, can be found in Effective Python by Slatkin [2].

3.3 Functions

We have already used a number of built-in functions and functions from external libraries, include the math.sin() and **range**() functions. In many situations, it is very helpful to write our own functions. Advantages of writing functions include the fact that it becomes easier to reuse the code you have written previously, and functions help to break our programs up into manageable pieces, which makes programming easier. The keyword **def** is used to define a function in Python. This keyword should be followed by the name of the function and variable names for any inputs. The set of instructions that make up the function appear in the block below the fist line. The construction of a function is illustrated through the following examples.

We want to write a function that will print out the area of a triangle given the size of the base and the height.

```python
def triangle(base,height):
    # two floating point inputs required:
    #   length of the triangle base
    #   height of the triangle
    area = 0.5*base*height
    # no return value
    print(area)

triangle(2,3)
```

Upon execution, this code should print "3.0" to the console output. The script first contains the code that defines the function. When run, the Python virtual machine loads the function but does not execute the function. It is not until the last line is reached that the function is called with two values that are passed into the function.

Of course having a function that just prints something to the screen after a calculation is probably not that useful. Instead, we should try to construct a function that returns the results of the calculations whenever possible. This can be illustrated by rewriting the function in the abovementioned example so that it returns the area (and then prints it to the screen).

```python
def triangle(base,height):
    # two floating point inputs required:
    #   length of the triangle base
    #   height of the triangle
    area = 0.5*base*height
    # one return value
    return area

size = triangle(2,3)
print(size)
```

Note the use of the keyword **return** at the end of the triangle function.

It can be helpful think of functions as virtual machines that take inputs, perform some operations on those inputs, and then return the results when they are finished. Figure 3.1 shows a visual representation of this process.

Step 1:
Build the function

Step 2:
Call the function

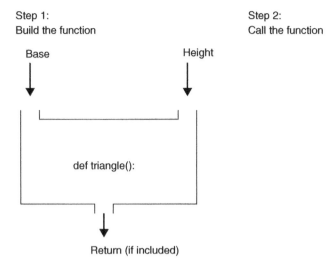

Base   Height

def triangle():

Return (if included)

Figure 3.1 A visual representation of the construction of the triangle function that acts as a machine that takes in two inputs, base and height, and returns the area of a triangle after performing the appropriate mathematical operations on the inputs.

The execution of the previous script starts by building a function. Then, the function is called in the second to last line and inputs are passed into the previously defined function. The return value from the function is stored in the variable $size$, which is printed to the screen in the final line.

Any variables declared inside a function are only defined within that specific function. A common mistake made by new Python programmers is to try to use a variable that was previously declared in a function somewhere outside the function. Some of the principles for accessing variables are illustrated in the example below.

The code below contains a global variable called everywhereVariable that is declared outside the function triangle, and a second variable, area, that is only defined within the scope of the function triangle.

```python
everywhereVariable = 102
def triangle(base=1,height=1):
    # two floating point inputs required:
    #    length of the triangle base
    #    height of the triangle
    area = 0.5*base*height
    area = 0.5*base*height
    print(everywhereVariable)
    # one return value
    return area
triArea = triangle(3)
print(triArea)
```

```
# next line gives an error
print (area)
```

Upon execution, this program should first define the variable `everywhere-Variable` and then it should load, but NOT execute the function triangle. Then, the function triangle should be executed when Python reaches the line where the function is first called (`triArea = triangle(3)`). The code should print "102" and "1.5" to the screen when the two print statements are executed. However, when we try to execute "**print**`(area)`" we should get an error message telling us that the variable "`area`" is not defined because the variable is only defined within the function – not outside the function.

One other Python feature to note in this example is the use of default values for the two input parameters of the triangle function. By setting default values, we can now call the function with or without values for `base` and `height`. If one value is passed into the function, that value is used for setting the first variable, (`base`). If a second value is passed into the function, both of the default values are replaced with the values passed into the function. Default values are very helpful in situations where parameters are unlikely to change in most situations.

The quadratic formula is used to find the roots of a quadratic polynomial of the form: $a \cdot x^2 + b \cdot x + c = 0$ where a, b, and c are given constants. The roots of the polynomial are

$$x = \frac{-b \pm \sqrt{b^2 - 4 \cdot a \cdot c}}{2 \cdot a} \tag{3.1}$$

and there may be up to two unique roots due to the $\pm$-sign. Further, if the term under the radical sign (i.e., square root sign) is negative, the roots are complex. We would like to develop a Python function that will calculates the roots of a quadratic polynomial given a, b, and c as inputs, assuming that the roots are real numbers, and it should print an error message if the roots are complex. An example function is shown below.

```
import math

def myquad(Ain, Bin, Cin):
    # check for complex roots
    if Bin**2 < 4.0*Ain*Cin:
        # if complex, print error
        print("Complex results not allowed")
        return (0,0)
    else:
        # if not complex, calculate roots
        x1 = -Bin + math.sqrt(Bin**2 - 4.0*Ain*Cin)
        x1 = x1 / (2.0*Ain)
        # calculate second root (plus/minus change)
        x2 = -Bin - math.sqrt(Bin**2 - 4.0*Ain*Cin)
```

```
        x2 = x2 / (2.0*Ain
        return (x1,x2)

# small program to call and test myquad()
a = 1.0; b = 4.0; c = 2.0
(out1, out2) = myquad(a,b,c)
print(out1, out2)
```

The myquad() function must be passed three numbers when it is called and these numbers are set to the variables Ain, Bin, Cin. The function then checks for complex roots using the term under the radical. If the roots are going to be complex, a warning is printed to the screen and zeros are returned for the values of the two roots. If the term under the radical is positive, the two roots, x1, x2, are returned as a tuple.

The final three lines in the code above allow one to test the myquad() function. Values are specified for the inputs to the function and then the function is called. Note that defining the function, which is done in the top half of the code, is not the same as calling or executing the function, which is done in the second to last line. You must call the function after it is defined if you actually want to run it. The call to the myquad() function in the second to last line has variables specified, out1, out2, in a tuple for receiving the return values. Alternatively, a single return variable could have been set and that variable would be set to a tuple with the two return values. For example, the function could have been called with the line out = myquad(a,b,c) and out would have been equal to a tuple with two values. If a tuple with three variables was used for receiving the returned values, that is, if the line was (out1, out2, out3) = myquad(a,b,c), a ValueError would result.

Potential changes to the myquad() function to make it easier to use include setting default values for the input variables, checking that the input variables are floating point numbers or integers using the type() function, and returning the complex roots for the case of a negative value under the radical. The interested reader is encouraged to modify the code above with these suggested changes.

As we will see later, it is often useful to combine related functions together into a single file with a *filename*.py extension. These functions can then be imported into other programs later on using **import** filename and called using filename.function. This is a great way to recycle code we have already written.

3.3.1 Pizza Example

The function below, given the number of people at a table, calculates the arc length for a slice of pizza if a single 16-inch diameter pizza is divided evenly among the people at the table and everyone receives just one slice.

```
import math
def arclength(numPeople):
    circumference = 16*math.pi
    if(numPeople < 1):
        print("Error: must have at least one person")
        return 0
    else:
        length = circumference/numPeople
        return length
```

```
print(arclength(6))
```

The code above should return 8.3776, indicating that each of the 6 people in the test problem should receive a slice with an arc length of 8.4 inches. You may want to try modifying the code so that it calculates the arc length when everyone receives more than one slice, but they all still receive the same number of slices.

3.3.2 Print Function

With the transition to Python version 3, the "print" command became a function, **print**(). The previous examples used basic forms of the print function, such as **print**(x) or **print**("hello world"). The print function is much more powerful and flexible than these simple examples, and this versatility can be very useful for producing more professional looking output from our programs. The examples below illustrate a few features of the print function.

```
>>> import math
>>> print("The value of pi is %lf" % math.pi)
The value of pi is 3.141593
>>> print("The value of pi is %le" % math.pi)
The value of pi is 3.141593e+00
>>> print("The value of pi is %d" % math.pi)
The value of pi is 3
>>> print("The value of pi is %.2lf" % math.pi)
The value of pi is 3.14
>>> print("The value of pi is %s" % "cherry")
The value of pi is cherry
print("Favorite pi? %.2lf or %s" % (math.pi, "cherry"))
Favorite pi? 3.14 or cherry
```

In the above example, the initial string in the print function contains formatting specifiers (e.g., %lf), and the string is followed by a percentage sign and then one or more variables (a tuple is used for multiple variables) to be printed in place of the format specifiers. Frequently used specifiers include floating point number, %f; long floating point number, %lf; long exponential number, %le; integer, %d; and string, %s. It is also possible in many cases to specify the number of digits that are printed using %*n.m*lf, where *n* is the number of digits before the decimal point and *m* is the number of digits after the decimal point. Both *n* and *m* are optional.

3.4   Debugging or Fixing Errors

Probably, the single greatest challenge that a novice programmer faces is correcting or fixing errors in their programs. The processes of correcting errors in computer code is commonly referred to as "debugging". The origin of the term debugging is frequently attributed to Grace Hopper, an early computer pioneer who discovered a moth stuck in a computer relay that was causing errors. The topic of debugging is very broad and a large number of books have been written that are dedicated to the topic of debugging. Experienced computer programmers almost always use debugger software to help with the process of finding errors. Debugging software allows programs to be executed one or a few steps at a time while the value of various variables can be continuously tracked. This ability to run a program in "slow motion" and with full variable exposure is very powerful. Debuggers for Python are included with most IDE's include the Spyder IDE. While the use of debuggers is not described here, a few important strategies for finding and correcting errors are listed below.

1) The single most common mistake made by novice programmers is trying to write 10 or more new lines of code without testing the code. Experienced programmers try to test their programs after writing just a few (2 to 5) new lines of code. Ideally, programmers like to start with a similar, working code that can be modified a few steps at a time to reach the desired code. *TEST OFTEN!!*
2) Print out the value of variables whenever possible during the initial writing of the code to ensure that the code is behaving properly. These "print()" statements are easy to remove later.
3) If a piece of code is not working, try to add a comment character, "#", before as many lines as possible. Hopefully, this will allow the code to run. Then, remove the comment characters one line at a time to find the line causing the problem.
4) Use the documentation available for the software libraries you are using to verify that the correct variables are being passed to functions in the library.

3.5   Top 10+ Python Error Messages

We end this chapter with a list of the most commonly encountered error messages and some common causes for those messages.

1) TypeError – this error is caused by trying to use a variable of one type in a situation that requires a different type. For example, trying to combine an "int" and a "str". The TypeError message is usually followed by the actual variable type followed by the required variable type.

2) IndexError – this error is caused by trying to access part of a list or array that is beyond the existing range. A frequent cause is forgetting that a list or array is indexed starting with zero. If an array (e.g., myarray) has 5 entries, then the 5th and final entry is accessed using an index of 4 (e.g., myarray[4]). In diagnosing the problem, it is often good to print the length of the array to the screen. The length of an list or container can be obtained using the `len()` function.

3) SyntaxError – this is caused by a violation of the Python syntax or formatting requirements. The most common cause is forgetting a colon (":") at the end of a line that requires one (e.g., lines starting with "if", "while", and "for")

4) SyntaxError: EOL while scanning string literal – this is a special syntax error that is usually caused by forgetting a quotation mark or using a mixture of single (') and double (") quotation marks.

5) NameError – happens when you try to use a variable that has not been defined. This frequently occurs when we forget to initialize a variable to a value.

6) ZeroDivisionError – probably, the easiest error message to understand, but it can be difficult to solve. It is always a good idea to print the values of variables to the screen to better understand when/how a variable is being set to zero instead of a nonzero value.

7) IndentationError – caused by inconsistent indentation in a block of code that should have been uniformly indented. Visual inspection can often reveal the problem unless the problem is caused by a mixture of "spaces" and "tabs". The solution is to not use "tabs" or use an editor that converts "tabs" into "spaces".

8) AttributeError – happens when we try to call a function that does not exist (or we misspell a function that does exist) in a library (e.g., math.sine() instead of math.sin()).

9) KeyError – only occurs with dictionaries when we use a key that does not exist.

10) SyntaxError: invalid syntax – can happen if we try to use one of the reserved Python keywords as a variable. The Python 3 keywords are: and, as, assert, break, class, continue, def, del, elif, else, except, False, finally, for, from, global, if, import, in, is, lambda, None, nonlocal, not, or, pass, raise, return, True, try, while, with, yield. Many chemical and biological engineers have struggled to fix this error when they tried to use "yield" as a variable name.

11) ValueError – frequently occurs when the wrong number of variables is specified for receiving the return values from a function.

Problems

3.1 The vapor pressure of a pure liquid, written p^*, is a strong function of temperature. To calculate the vapor pressure at a given temperature, T, it is common to use Antoine's equation:

$$\log_{10} p^* = A - \frac{B}{T + C}, \tag{3.2}$$

where A, B, and C are constants that can be looked up for different liquids. Write a function that has A, B, C, and T (in $^\circ$C) as inputs and returns the vapor pressure, p^*. Hint: $10^{\log_{10} x} = x$.

3.2 Starting with Antoine's equation (see previous problem), write a function that has A, B, C, and p^* (in mm Hg) as inputs and returns the temperature, T, for the given vapor pressure, p^*.

3.3 Write a function that compares items sequentially between two lists, for example, list a and list b, and calculates the total number of times that the item in list a is larger than the item in list b. If the lists are of different length, the comparison should only be performed for the total items in the shorter list (i.e., ignore items in the longer list that are beyond the end of the shorter list). The Python function **len** () is very useful for getting the length of a list as an integer.

3.4 Write a function called FtoC(T) that receives a temperature in Fahrenheit as the input and returns the temperature in Celsius as the return value. Write a second function called CtoF(T) that does the opposite – receives a temperature in Celsius as the input and returns the temperature in Fahrenheit. Demonstrate one of the functions by inputting the current temperature at the location of your birth using the standard temperature measurement unit at that location and print out the temperature in the other system of units. For example, I was born in Bozeman, Montana, USA, so my input would be the current temperature in Fahrenheit, and I would print out the temperature in Celsius.

3.5 Write a script that stores each line of the song "Happy Birthday" as a separate string in a list. Then, input from the user a number corresponding to the number of lines that they would like printed to the screen. Check the number to determine that it is valid before printing the lines to the screen.

3.6 Write a function that receives a single word (i.e., a string of text) as input and then prints out that word in Pig Latin. For anyone unfamiliar with Pig Latin, this requires moving the first letter to the end and then adding "ay".

A couple hints: (1) specific letters within a string can be accessed just like a vector of numbers – for example, if myString = "jeff", then myString[0] returns "j" and myString[2:3] returns "ff", and (2) you can concatenate strings with the plus sign, so myString + "heys" gives "jeffheys". Finally, write the function so that it can receive a sentence as the input and then translate each word – hint: use the string.split() function.

3.7 Write a function that receives a string as the only input. The function should then iterate through each letter in the string (note that the length of the string can be obtained using `len()` and the individual characters for string a can be accessed using `a[i]`) and count the number of each type of vowel, a, e, i, o, or u. The function should also count the vowel regardless of case: both upper and lower case vowels should be counted (hint: character *x* or string *x* can be forced to be lower case using *x*.lower(), which makes counting easier). The function should return five integers, the number of vowels of each type, a, e, i, o, and u. The function should be tested using the string "Alphabet" as the input.

3.8 The decision tree shown below (Figure 3.2) has been developed by Chaste Bank after they analyzed the probability of prepayment on each mortgage they have issued over the past few years. You have been hired by Chaste to implement the decision tree in Python so that the answers to the questions can be entered by a bank representative to determine the probability of prepayments. The code should ask "yes" or "no" questions and accept "Y", "y", "N", or "n" as the answer (i.e., you will want to convert the string to

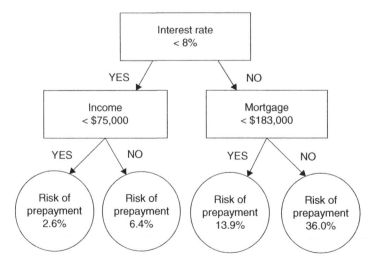

Figure 3.2 A decision tree for assessing prepayment risk on a mortgage [3].

upper- or lower case to ensure that the case is known). The code should then print the prepayment probability to the screen.

3.9 You are planning to purchase a new car and you are suddenly concerned about the annual operating costs. Your list of car choices has been narrowed down to

Car (2015 model year)	mpg
Chevrolet Silverado	15.0
Chevrolet Corvette	20.0
Chevrolet Spark	30.0

where the fuel efficiency (in miles per gallon) was obtained from www .fueleconomy.gov/feg/pdfs/guides/FEG2015.pdf. Create a Python script that stores the model name (e.g., "Silverado") and fuel efficiency for each vehicle in two separate lists. Then, loop through the cars and calculate the annual operating cost using the current price of gas at a station near you and assuming that you will drive 10,000 miles per year. The program should print the annual operating cost for each vehicle to the screen as well as the model name of the vehicle.

Additional Resources

Reference books on programming in Python:

- Python Pocket Reference by Lutz [4]
- Python Programming by Zelle [5]
- Introduction to Computing Using Python by Perkovic [6]

Books on numerical programming in Python:

- Numerical Methods in Engineering with Python by Kiusalaas [7]
- Python Scripting for Computational Science by Langtangen [8]

References

1 van Rossum, G., Warsaw, B., and Coghlan, N. (2015) *Style Guide for Python Code*, http://www.python.org/dev/peps/pep-0008/.

2 Slatkin, B. (2015) *Effective Python. 59 Specific Ways to Write Better Python*, Addison-Wesley, Upper Saddle River, NJ.

3 Siegel, E. (2013) *Predictive Analytics*, John Wiley & Sons, Inc., Hoboken, NJ.

4 Lutz, M. (2014) *Python Pocket Reference*, O'Reilly Media, Inc., Sebastopol, CA, 5th edn.

5 Zelle, J. (2010) *Python Programming: An Introduction to Computer Science*, Franklin, Beedle & Associates Inc., Portland, OR, 2nd edn.

6 Perkovic, L. (2015) *Introduction to Computing Using Python: An Application Development Focus*, John Wiley & Sons, Inc., Hoboken, NJ, 2nd edn.

7 Kiusalaas, J. (2010) *Numerical Methods in Engineering with Python*, Cambridge Press, New York, NY, 2nd edn.

8 Langtangen, H.P. (2010) *Python Scripting for Computational Science, Texts in Computational Science and Engineering*, Springer-Verlag, Berlin, 3rd edn.

4

External Libraries for Engineering

Now that we have covered the basics of programming in Python over the past two chapters, including variable assignments, functions, loops, and conditionals, we are now ready to examine some external libraries that are used frequently in engineering and scientific computing. The exploration will begin with numpy, which forms the foundation of numerical array storage in Python. Numpy forms the foundation of much of the Scipy library and matplotlib library for plotting. The use of Matplotlib for basic plotting is also covered in this chapter.

4.1 Numpy Library

The numpy library adds powerful linear algebra data structures to Python. It allows us to construct and manipulate vectors and tensors very efficiently, and it is also widely used by other libraries that provide, for example, plotting, data science algorithms, and linear algebra solvers. The goal of this section is to provide a very brief introduction to a few important features of numpy. Throughout the rest of the book, additional features and options will be demonstrated. The online documentation and tutorials for numpy are also a very valuable source of information about the numpy library.

4.1.1 Array and Vector Creation

There are a number of different interfaces provided by numpy for constructing vectors and arrays. The simplest approach is to simply pass the numpy.array() method a Python list:

```
myvector = numpy.array([5,3,7])
myarray = numpy.array([[2,3],[6.7,1.0]])
print(myvector)
print(myarray)
print(myvector.dtype)
print(myarray.dtype)
```

Chemical and Biomedical Engineering Calculations Using Python®, First Edition. Jeffrey J. Heys.
© 2017 John Wiley & Sons, Inc. Published 2017 by John Wiley & Sons, Inc.
Companion Website: www.wiley.com/go/heys/engineeringcalculations_python

and the output is

```
[5 3 7]
[[ 2.    3. ]
 [ 6.7  1. ]]
int64
float64
```

If a single, one-dimensional list is passed into the numpy.array() method, a one-dimensional vector is created. If a nested list is passed into the numpy.array() method, a two-dimensional tensor is created. Every numpy array has a "data type" or "dtype" and this parameter specifies both the numerical type and precision of the values stored in the array. In the previous example, the first array, "myvector", was constructed using a list of integers so numpy automatically set the dtype as "int64" or 64-bit integers. The second array, "myarray", was constructed using a mixture of integers and floating point numbers so the dtype was set as "float64" or 64-bit floating point numbers.

In engineering calculations, we typically want to work with floating point numbers and floating point arithmatic, so it is desirable to specify the data type when it is first created. This can be accomplished by setting the dtype parameter at construction. The construction of "myvector" in the previous example can be slightly modified to specify the desired dtype:

```
myvector = numpy.array([5,3,7], dtype=numpy.float64)
print(myvector)
print(myarray.dtype)
```

and now the output is:

```
[ 5.   3.   7.]
float64
```

Note the decimal points after the values in the vector, which denote floating point values instead of integers. The integers in the list used for construction were converted to floating point values.

Using a list for array construction works great for very small arrays where we already know the values. In practice, however, we will usually construct an array of zeros of the size we want and then use loops to insert the desired values into the array. This approach is illustrated in the following example:

```
import numpy

size = 3
myarray = numpy.zeros((size,size))

for i in range(size):
    for j in range(size):
        myarray[i,j]=1.0/(i*j+1.0)

print(myarray)
```

The program constructs a square, two-dimensional array to the size specified by the first argument passed into the `numpy.zeros()` function. If a vector or one-dimensional array of zeros is desired, then the function call can be `numpy.zeros(size)`. If a two- or higher-dimensional array is desired, a tuple should be passed into the `numpy.zeros()` function as is done in the example above. The zeros within the array are then replaced by the values calculated inside the nested loop. This approach to first allocating space for an array and then overwriting the initial values is typically more computationally efficient than first constructing an empty array and then appending values.

We need to make a very important observation from the previous example: *numpy arrays are the same as other Python containers (lists, tuples, etc.) and are indexed starting with zero!* In other words, if we have a numpy vector that contains 5 numbers (i.e., it has length 5), those numbers are accessed with the indices 0, 1, 2, 3, and 4. Notice how if we loop from 0 to a number less than the size of the vector (in this case we loop from 0 to 4), we loop through all the indices without including 5. This observation is important, and forgetting how numpy arrays are indexed leads to many troublesome bugs in the code.

It is sometimes helpful in array construction to build an array where the values increase incrementally. For example, we might wish to construct an array of length 10 that contains the integers from 1 to 10. One good feature of such a vector is that it also allows us to demonstrate how to access a subsection of the vector using array slicing

```python
import numpy

myarray = numpy.arange(1,11)
print(myarray)
myarray[3:6] = numpy.array([300,400,500])
print(myarray)
```

The code in this example begins by constructing a vector from 1 to 10, but then, we replace 3 of the values within the vector with much larger values. Notice how we replace the values stored at indices 3, 4, and 5 because the slicing command – 3:6 – does not include the last index (6) in the slice. Also recall the first entry in the vector (in this case "1") is at index zero. Therefore, index 3 initially contains the value 4, which is replaced with 300. Then, the value at index 4, which is initially 5, is replaced with 400. The result of running the example code should be

```
[ 1  2  3  4  5  6  7  8  9 10]
[ 1  2  3 300 400 500  7  8  9 10]
```

More general functions for constructing numpy arrays that consist of floating point numbers that vary between two end values are the `numpy.linspace()` and `numpy.logspace()` functions. The two functions are demonstrated below.

```
import numpy

lin = numpy.linspace(1.0,3.0,6)
print(lin)
logger = numpy.logspace(1.0,3.0,num=5)
print(logger)
```

The `linspace()` function requires a minimum of two numbers as inputs: a starting value and a stopping value. The function creates a numpy vector with a starting value as the first number and a final stopping value as the second number. The entries in the vector between the starting and stopping values are linearly (or evenly) spaced and the total number of entries in the vector can be specified using a third number passed to the function (the default is a vector of length 50). The vector constructed by the `linspace(1.0, 3.0, 6)` function above is `[1. 1.4 1.8 2.2 2.6 3. ]`. The `logspace()` function is very similar, but instead of specifying a starting and a stopping value, the exponent for the starting and stopping values are specified. The default base is 10, so the `logspace(1.0, 3.0, num=5)` call above creates a vector starting at 10^1 and ending at 10^3 with length 5. The intermediate entries in the vector are not linearly spaced but are based on linearly spaced exponents. As a result, the `logspace()` function above creates the numpy vector: `[ 10. 31.62 100. 316.2 1000. ]`.

Fibonacci Sequence

The Fibonacci sequence is an important mathematical series that is frequently found in nature (e.g., the arrangement of sunflower seeds). The first two values in the sequence are defined as 0 and 1. Additional terms in the sequence are calculated by summing the previous two terms in the sequence. The function below stores the Fibonacci sequence in a numpy array.

```
import numpy

seqLength = 10
seq = numpy.zeros(seqLength,dtype=numpy.int32)

seq[0]=0
seq[1]=1
for i in range(2,seqLength):
    seq[i]=seq[i-1]+seq[i-2]

print("Final sequence: ",seq)
```

4.1.2 Array Operations

To review, we have discussed the construction of numpy vectors and arrays, and we have discussed how to access the various entries within the vectors and arrays. Now let us explore a few different operations that can be performed on matrices and vectors with the following example.

Try running the code below:

```python
import numpy

myarray = numpy.arange(5)
print(myarray) # output: [0 1 2 3 4]
print(myarray.shape) # output: (5,)
myarray = myarray*4
print(myarray) # output: [0 4 8 12 16]
yourarray = numpy.ones(5)
theirarray = myarray - 3*yourarray
print(theirarray) # output: [ -3. 1. 5. 9. 13.]
print(numpy.dot(myarray,theirarray)) # 360.0
itsarray = numpy.outer(myarray,theirarray)
print(itsarray)
print(itsarray.shape) # output: (5,5)
```

This code segment begins by building a sequential vector of length 5. The "shape" property of that vector should return "(5,)", which tells us that the array has 5 rows and no additional columns. This initial vector is then multiplied by 4 and a second and a third array are built. The third array, called "theirarray," is actually built by taking the first array, "myarray," and subtracting an array of 3s from it. Note that adding or subtracting one array from another requires that the arrays have the same size and shape – that is why it is often a good idea to print the value of the size and shape to the screen so that you can confirm that the sizes are the same. The example continues by taking a dot product of two vectors (they must have the same length) and an outer product.

This example barely scratched the surface of what is possible with numpy. For example, we have not yet covered matrix inversion or eigenvalue calculations, but many of these topics will be discussed in later chapters.

4.1.3 Getting Helping with Numpy

One function that we briefly saw earlier is numpy.logspace(), and it might be helpful to learn a little more about this function. Begin by using the help() function to read the description:

```
>>> help(numpy.logspace)
Help on function logspace in module numpy.core...
```

```
logspace(start, stop, num=50, endpoint=True, base=10.0)
    Return numbers spaced evenly on a log scale.
...
```

Most of the help document was trimmed here to save space. The `start` and `stop` values that are passed into this function represent the exponent for the actual starting and stopping values, that is, the actual starting value for the array is $base^{start}$ or 10^{start} if the default base value of 10.0 is used. The actual ending value for the array is $base^{stop}$. An example of building an array from 0.1 to 100.0 of length 8 is as follows:

```
>>> import numpy
>>> print(numpy.logspace(-1,2,8))
[  0.1   0.26826958   0.71968567   1.93069773   5.17947468
13.89495494   37.2759372   100. ]
```

4.1.4 Numpy Mathematical Functions

Numpy also has built in functions for performing mathematical operations on numpy vectors and arrays. For example, the following Python script:

```
import numpy
import math

g = numpy.arange(2,4,0.5)
print(g)
h = math.sin(g)
print(h)
```

will generate an error because the sin() function in the math library is expecting a single number and not an array of numbers. Instead, the code should be written using the sin() function that is part of numpy:

```
import numpy

g = numpy.arange(2,4,0.5)
print(g)
h = numpy.sin(g)
print(h)
j = numpy.power(g,2.5)
print(j)
```

which gives the output:

```
[ 2.   2.5  3.   3.5]
[ 0.90929743   0.59847214   0.14112001  -0.35078323]
[  5.65685425   9.88211769   15.58845727   22.91765149]
```

In the abovementioned example, the `numpy.power()` function is not strictly necessary for raising every number in the vector to the power 2.5, and, instead, we could have just used `g**2.5`.

4.1.5   Random Vectors with Numpy

The numpy library contains functions for generating arrays of random numbers. These functions are in numpy.random, the random sampling part of the library. Some of the most useful functions are demonstrated below.

```
import numpy as np

a = np.random.randint(0,10, size = 5)
print(a)
b = np.random.random(size=(2,2))
print(b)
print(b*10.0)
c=np.arange(5)
np.random.shuffle(c)
print(c)
```

The numpy.random.randint(low,high,size) function returns a numpy array of the specified size (a vector of length 5 for the example above) with the random integers drawn from between the low and (high − 1), that is, all the random integers will be greater than or equal to the low value and strictly less than the high value. Random integers can and do repeat. For example, the function above generated: [3 0 0 9 8] on one occasion. The numpy.random.random(size) function generates random floating point numbers greater than or equal to zero and strictly less than 1.0. The example above generates a 2 × 2 array of random floating point numbers. If random floating point numbers larger or smaller than 1.0 are desired, they can be obtained by multiplying a random array by a scaler. In the example above, random floating point numbers between [0, 10.0) are obtained simply by multiplying the array by 10.0. The final function discussed here is the shuffle function. The numpy.random.shuffle() function will simply shuffle the locations of values in a numpy vector or array. Note that a new array is not returned, but, instead, the original array passed into the function is forever shuffled. In the example above, a sequential array of integers is shuffled. The original array of [0 1 2 3 4] was shuffled into [0 4 2 1 3], in one instance.

4.1.6   Sorting and Searching

The sorting of numpy vectors can be performed using the numpy.sort() function. This is demonstrated in the script below where a vector of random floating point numbers is generated and then sorted.

```
import numpy as np

a = np.random.random()
print(a)
b = np.sort(a)
print(b)
```

The results of running this script are different every time it is run due to the generation of random numbers, but the output from one example run is shown below.

```
[ 0.46118278   0.00577888   0.95539835   0.652181   ]
[ 0.00577888   0.46118278   0.652181     0.95539835]
```

Sorting two or higher dimensional arrays is also possible. Typically, the axis for sorting is specified when the `sort()` function is called.

In engineering, we are often interested in the maximum or minimum value from a list of values. For example, we might want to know the maximum stress in a fluid or structural beam because that value often needs to remain below some threshold. Two different pairs of function are useful for locating and obtaining extreme values. The first pair: `numpy.amax()` and `numpy.argmax()` return the maximum value in a numpy vector and the index or location of that maximum value, respectively. These two functions are used in the example below to find the maximum value and its index for a vector of random values.

```python
import numpy as np

a = np.random.random(4)
print(a)
print(np.amax(a))
print(np.argmax(a))
```

and the output from one random run was

```
[ 0.75992421   0.47808563   0.9004158    0.26679892]
0.900415801936
2
```

Since numpy vectors are indexed starting with 0, the maximum value occurs at index 2 or the third value in the vector. The minimum value and index of the minimum value may be found using `numpy.amin()` and `numpy.argmin()`, and they have an identical input/output format as the maximum value functions.

One final function for searching numpy vectors that is frequently useful is the `numpy.nonzero()` function, which returns the index of all elements (or entries) in a numpy vector that are nonzero. The length of the returned array corresponds to the total number of nonzero values.

4.1.7 Polynomials

Polynomials are represented on computers by storing the coefficients of the different terms in the polynomial within a vector. Numpy includes some functions for building, managing, manipulating, and evaluating polynomials. The first rule that we need to recognize, however, is that in numpy (and most other computational tools that support polynomials), the polynomial must be written with the zero-order term first and then progressing sequentially to higher-order terms. For example, if we want to use the polynomial:

$$f(x) = x^2 - 2x - 3,$$

we must begin by reordering the terms in the polynomial as

$$f(x) = -3 - 2 \cdot x + 1 \cdot x^2.$$

This polynomial can be represented in numpy using the `numpy.polynomial` package. The use of this package to build and evaluate the example polynomial from above is demonstrated below.

```
import numpy.polynomial as np

f = np.Polynomial([-3., -2., 1.])
print(f.roots())
(x,y) = f.linspace(8,domain=[-2,5])
```

The variable f holds the coefficients of the polynomial, starting with the zero-order term. The polynomial package includes a `roots()` function that returns the roots of the polynomial (in this case, the function returns `[-1. 3.]`). Finally, the `linspace()` function evaluates the polynomial at 8 points in the domain $x \in [-2, 5]$. The `linspace()` function returns two arrays, one containing the x-values and the other containing the y-values resulting from the evaluation of the polynomial. These arrays can be plotted to visualize the polynomial as shown in Figure 4.1 and described in Section 4.2.

4.1.8 Loading and Saving Arrays

Numpy includes extensive support for writing vectors and arrays into files and then loading those files at another time. Here, we are only going to explore the reading and writing of files using numpy's binary format and the traditional ASCII text format. The use of both formats is illustrated in the script below.

Figure 4.1 A figure of the polynomial $f = -3 - 2x + 1$. Polynomial evaluation used the numpy polynomial package, and the plot was generated using matplotlib.

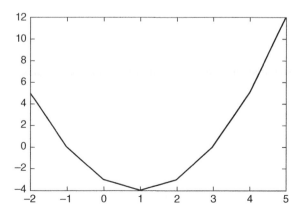

```
import numpy as np

x = np.arange(5)
print(x)
np.save('binFile',x) # .npy extension auto added
y=np.load('binFile.npy') # binary file load
print(2.0*y)
np.savetxt('txtFile.out',x)   # extension required
z = np.loadtxt('txtFile.out')
print()
```

The script begins by creating a vector of five sequential integers. The vector is then saved to a binary file with the filename *binFile.npy* using the numpy.save() function and passing the function the filename and the array variable name. Note that the numpy.save() function automatically adds the proper filename extension. The binary file is read using the np.load() function (note that the correct file extension is required), and the array is set to a new variable name. The second half of the script repeats this same process but the numpy.savetxt() function is used to save the original array to an ASCII text file. The same file is then read and the array stored to a new variable.

One obvious question from this example is: which storage format is better, binary or text? The binary storage format is more computationally efficient. The file is smaller and reading or writing large arrays is faster. The text file storage format is easier for humans. The values in the file can be read and edited by a huge array of software packages, include MATLAB, Excel, and others. For most applications, text files are simpler and the better choice. Only in cases of very large arrays should binary files be used.

4.2   Matplotlib Library

Matplotlib is one of many libraries that adds plotting capabilities to Python. It is a particularly good choice here because it is well integrated with numpy and it provides a relatively high-quality output. To use Matplotlib within a Python script, it is recommended that the user import matplotlib.pyplot. It can become cumbersome to type matplotlib.pyplot repeatedly in our codes, so consider using the command: **import** matplotlib.pyplot as plt, which allows functions within the plotting library to be called using the "plt.function()" format. Matplotlib also support the "pylab" interface, which can be imported as **import** pylab. The two interfaces are nearly identical and both provide plotting functionality and an interface that is similar to MATLAB. The pylab interface is older so it is recommended that users move toward the pyplot interface.

Let us begin with a simple example. The following script builds a vector with 100 entries that span from 0 to 10.0 using numpy's linspace() function. Then, pyplot is used to generate a scatter plot, where x is from 0 to 10 and y is $\cos(x)$.

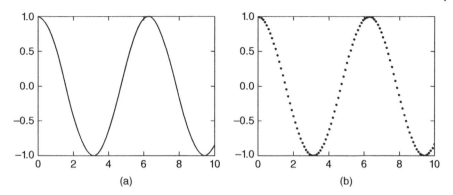

Figure 4.2 A figure of a cos()-wave generated using pylab with (a) a solid line (and (b) circles.

```python
import numpy
import matplotlib.pyplot as plt

x=numpy.linspace(0,10, num=100)
y = numpy.cos(x)
plt.plot(x,y)
plt.show()
```

The resulting figure is shown in Figure 4.2(a). The final line of example code (plt.show()) is required because it causes the figure to persist on the screen and become interactive. If this line is omitted, the figures may never be plotted to the screen or is plotted for a fraction of a second before it disappears on some operating systems. Think of the plt.show() function as causing the program to pause and wait for the user to decided what they want to do with the plot. The plt.show() function is not required on all operating systems.

If the function call to plt.plot(x,y) is replaced with plt.plot(x,y, 'bo'), the plot is constructed with circles instead of a solid line, as shown in Figure 4.2(b).

It is also possible to place multiple curves on the same plot and include axis labels and figure titles as illustrated in the following example of polynomial curves.

```python
import matplotlib.pyplot as plt
import numpy as np

x = np.linspace(0, 2, 200)

plt.plot(x, x, label='linear')
plt.plot(x, x**2, '.', label='quadratic')
plt.plot(x, x**3, '--', label='cubic')

plt.xlabel('x-axis label')
plt.ylabel('y-axis label')

plt.title("Polynomials")
```

```
plt.legend()

plt.show()
```

This example illustrates how every call to functions in the matplotlib library are applied to the current, active figure. This behavior is similar to Matlab. Notice that axis labels are added to the figure using the `plt.xlabel()` and `plt.ylabel()` function calls with the desired string of text being passed into the function. A title is added to the figure using `plt.title()` function call, but, in common engineering practice, titles are not included for figures. In the future, we will use `plt.figure()` in the matplotlib library to generate additional figures and avoid having everything on the same figure. The result of running the example code is shown in Figure 4.3.

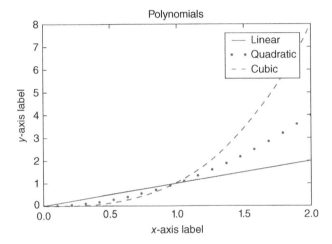

Figure 4.3 A plot showing a linear, quadratic, and cubic polynomial all on the same plot with a legend identifying each curve.

Before ending this section, the power of Matplotlib is illustrated through a slightly more complex example that shows a contour plot of a function with two independent variables.

```
import pylab
import numpy

def f(x,y):
    return (1-x/2+x**2+y**3)*numpy.exp(-x**2-y**2)

n = 256
x = numpy.linspace(-2,4,n)
y = numpy.linspace(-2,4,n)
X,Y = numpy.meshgrid(x,y)
```

```
C = pylab.contour(X, Y, f(X,Y), 8)
pylab.clabel(C,inline=1)
pylab.colorbar(C,orientation='vertical')
pylab.show()
```

The function that is plotted is defined by the function $f(x,y)$. The independent variables, $x \in (-2,4)$ and $y \in (-2,4)$, are stored in vectors created using numpy's linspace() function. The numpy meshgrid function extends the one-dimensional vectors over a two-dimensional array. The contour plot consists of eight contour lines, which are labeled, and a colorbar is added to the right of the plot. Color bars are largely unnecessary and unattractive for this style of contour plot, but one is included here to illustrate the simplicity with which it can be added. The resulting figure is shown in Figure 4.4.

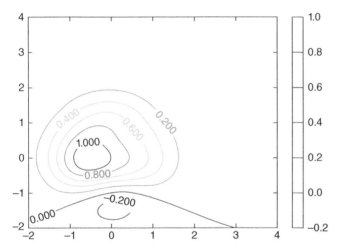

Figure 4.4 A contour plot of a function, $f(x,y) = (1 - x/2 + x^2 + y^3) \exp(-x^2 - y^2)$ for $x \in (-2,4)$ and $y \in (-2,4)$.

Matplotlib is a comprehensive plotting library, large enough that an entire book has been written to document all of the many different style of figures and options. For more information about the library as well as documentation describing the interfaces into the library, see the tutorials posted on the Matplotlib library website: matplotlib.org.

4.3 Application: Gillespie Algorithm

This chapter introduced two important libraries for solving chemical and biological engineering problems using Python. Before ending this chapter, let us explore the use of these two libraries when implementing an important algorithm in modeling biochemical reactions: the Gillespie algorithm.

Classical chemical reaction kinetic models are derived by assuming that a system has 10^6 or more well-mixed molecules. When one remembers that one mole of a material has more than 10^{23} molecules, the assumption that a reactor contains more than 10^6 molecules is almost always valid. However, if we want to model biochemical reactions in a single cell, the number of reacting molecules present is often on the order of 100 or 1000 molecules. Hence, using classical chemical reaction kinetics to model reactions within a cell may not be valid. The Gillespie algorithm, in contrast, is not based on the assumption of a large number of molecules. Instead, the algorithm is stochastic and based on tracking a discrete number of molecules. Briefly, the algorithm is based on the generation of random numbers (i.e., analogous to rolling a dice) for two calculations: (1) using the reaction rate and a random number, the algorithm determines whether or not a reaction occurred for a random molecular collision and (2) using a random number to discretely approximate the time until the next collision. Detailed derivation and description of the algorithm are beyond the scope of this brief presentation, but the interested reader is referred to the original Gillespie paper [1] or any of the thousands of papers on the algorithm written in the past few decades. A Python library for modeling discrete, stochastic reactions using a number of different algorithms, including the Gillespie algorithm, is StochPy (stochpy.sourceforge.net).

To illustrate the Gillespie algorithm, consider the chemical reaction

$$A \underset{k_2}{\overset{k_1}{\rightleftharpoons}} B,$$

where k_1 is the forward rate constant (units are per time) and k_2 the backward rate constant (same units). Classical chemical reaction kinetics would predict that this reaction will proceed toward equilibrium and behave like a first-order reaction. Alternatively, the Gillespie algorithm for this reaction is:

```python
import numpy
import matplotlib.pyplot as plt

k1 = 1.0 # forward rate constant
k2 = 0.1 # reverse rate constant

maxReact = 1000 # maximum number of reactions
numMol = numpy.zeros((2,maxReact),dtype = numpy.float)
timePt = numpy.zeros(maxReact, dtype = numpy.float)
numMol[0,0] = 175    # initial number of A's
numMol[1,0] = 25     # initial number of B's
timePt[0] = 0.0      # initial time

rands = numpy.random.rand(2,maxReact)

for i in range(maxReact-1):
```

```
    proB = k1*numMol[0,i] # probability of forming B
    proA = k2*numMol[1,i] # probability of forming A
    # calculate time till next reaction
    dt = -numpy.log(rands[0,i])/(proB+proA)
    timePt[i+1] = timePt[i] + dt
    if rands[1,i] < (proB/(proA+proB)): # form B?
        numMol[0,i+1] = numMol[0,i] - 1.0
        numMol[1,i+1] = numMol[1,i] + 1.0
    else:                      # else we form A
        numMol[0,i+1] = numMol[0,i] + 1.0
        numMol[1,i+1] = numMol[1,i] - 1.0

plt.plot(timePt,numMol[0,:], label="A")
plt.plot(timePt,numMol[1,:], label="B")
plt.xlabel('time')
plt.ylabel('number of molecules')
plt.legend()
plt.savefig('gillespie.png')
```

The first 20 lines of the algorithm are setting up the problem to be solved. The script begins by loading the numpy and matplotlib libraries and then the rate constants for the reaction are specified. Note that the forward rate is much faster than the reverse rate so we should have more B than A at equilibrium. The Gillespie algorithm is based on counting the exact number of molecules in the system for a preset number of reactions. In this case, we choose to simulate 1000 reactions and an empty numpy array, numMol, is constructed to later store the number of molecules of A (column 0) and molecules of B (column 1) after each of the 1000 reactions. A second empty numpy array is constructed to store the time at which each reaction occurs. Recall that the time between reactions is stochastic – that is, a role of the dice. The setup phase ends with the construction of a numpy array full of random numbers. The array has two columns, one for each reaction.

After the setup phase, the main part of the algorithm is an iterative for loop that contains the calculations for each of the 1000 reactions. Each iteration begins by calculating the probability of the forward reaction (forming B) and the reverse reaction probability (forming A). The time till the next reaction is also calculated and using a random value and the reaction probabilities, that is, smaller probabilities imply more time till the next possible reaction. On the basis of the probabilities of forming A and B, the number of molecules of A and B is updated in the numMol array. Finally, after stochastically simulating 1000 reactions, matplotlib can be used to plot the number of molecules of A and B in the system as a function of time. While every simulation result is different due to the stochastic nature of the algorithm, an example of a single simulation is shown in Figure 4.5. The curve resembles a noisy version of a classical first-order reaction.

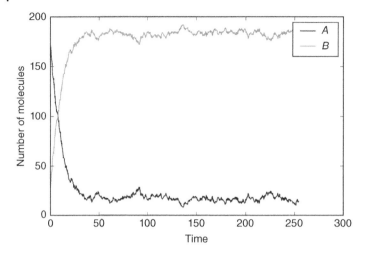

Figure 4.5 Number of molecules of A and B for a first-order equilibrium reaction simulation using the Gillespie algorithm. Forward rate is $k_1 = 1.0$ and the reverse rate is $k_2 = 0.1$ (both per time).

Problems

4.1 You have been hired by NASA to develop a short Python script that when executed, asks the user to input their weight on earth and then select the planet that they are currently occupying. The program should then calculate and print out to the screen the individual's weight on their current planet. Use a relative weight table such as that reproduced below to simplify the calculation. Three notes: (1) the input() function in Python returns a string, which you will need to convert into an integer using the int() function, (2) it is easiest to ask the user to input a number for the planet selection (i.e., enter 1 for Mercury, 2 for Venus, etc.), and (3) you should check the planet selection using an "if" statement to ensure that it is valid.

Mercury	0.38
Venus	0.91
Earth	1.0
Mars	0.38
Jupiter	2.34
Saturn	1.06
Uranus	0.92
Neptune	1.19

4.2   You have been hired as a consulting engineering to answer a question from the Engineering Department at Mosure University. The department would like to determine the probability of two students in the same class having the same birth date. We are only considering the day of the year (e.g., February 3) and not the year of birth. Warning, you probably will not believe the result.

Because we are engineers and not statisticians, you are being hired to develop a computer program that is capable of generating students with random birth dates for a specified class size. The program will simulate a specified number of classes and determine the fraction of classes that contain at least two students with the same birthday.

The program should have the following attributes:

a) The user can set a variable "NumTrials" equal to the number of virtual classes they wish to simulate. I initially set this variable to 1000.

b) The user can set a variable "ClassSize" equal to the size of the class (e.g., 40, 60, 80, and 200).

c) You should write a function that receives as input a numpy vector of length ClassSize that contains the birthday for every member of the class. The function should then determine if any two students have the same birthday (and return 1 or True in this case) or not (and return 0 or False in this case).

d) The program should run multiple trials. For each trial, you should construct a numpy vector of length ClassSize containing randomly assigned birthdays. I recommend using a number between 0 and 364 to represent the birthday, and I recommend using the numpy.random.randint() function to construct the vector of random birthdays.

Your final report to the department as a consultant should consist of a discussion of your findings and, most importantly, a plot of the probability of a class containing two students with the same birthday versus class size for classes between 20 and 200 students. Note that you should not simulate every possible class size between 20 and 200, just pick 5 or 6 class sizes over that range. The report should be in the form of a memo to the department from you. Hint: in a class of 40 students, there is a 89% probability of two students having the same birthday – see, I told you that you would not believe the result.

One final warning – you will need to calculate the probability, that is, the fraction of trials where two students had the same birthday divided by the total number of trials. Calculating this fraction is easy, I used "successes/NumTrials". However, both of these variables were integers (e.g., 891/1000), and Python 2.x reports the result as an integer (i.e., 0 or 1) instead of a fraction (i.e., 0.891). To ensure a floating point result, I used "(1.0*successes)/NumTrials".

4.3 The "Monty Hall Problem" poses following challenge: imagine you are on a game show and are faced with three doors. Behind one of the doors is a great prize, and behind the other two doors is something of little value. The game show host asks you to pick one of the doors, which you do. On the basis of your choice, the game show host open one of the other two doors and reveals something of little value. At this point, two doors remain closed, one of them concealing the prize. The game show host offers to let you switch your choice of doors. You fear that the host is trying to trick you. Should you switch.

Determine the answer to this question by writing a Python algorithm that can simulate a large number of virtual games (e.g., $n = 10,000$ games). Create a numpy array that holds a random integer: 1, 2, or 3, corresponding to the winning door for each game. An easy way to build this array is the function: `numpy.random.randint(1,4,n)`. Without loss of generality, it is possible to have the contestant select the same door every game for all n games. Next, allow for the selection of whether to switch the door selected or not. On the basis of the door selection and the switch selection ("yes" or "no"), the algorithm should be able to iterate through the n games and determine the fraction won. Does the fraction won depend on whether you switch or not?

Additional Resources

Recommended books on the external libraries covered here:

- Matplotlib for Python Developers by Tosi [2]
- Learning SciPy for Numerical and Scietific Computing [3]
- High-Performance Python by Gorelick and Ozsvald [4]

Recommended book on linear algebra:

- Introduction to Linear Algebra by Strang [5]

References

1 Gillesphie, D.T. (1977) Exact stochastic simulation of coupled chemical reactions. *J. Phys. Chem.*, **81**, 2340–2361.

2 Tosi, S. (2009) *Matplotlib for Python Developers*, Packt Publishing, Birmingham, UK.

3 Rojas, S.J., Christensen, E.A., and Blanco-Silva, F.J. (2015) *Learning SciPy for Numerical and Scientific Computing*, Packt Publishing Ltd., Birmingham, UK, 2nd edn.

4 Gorelick, M. and Ozsvald, I. (2014) *High Performance Python: Practical Performant Programming for Humans*, O'Reilly Media, Inc., Sebastopol, CA, 1st edn.

5 Strang, G. (2009) *Introduction to Linear Algebra*, Wellesley-Cambridge Press, Wellesley, MA, 4th edn.

5

Symbolic Mathematics

5.1   Introduction

When we have a mathematical equation or equations that describe some phenomena, there are basically two approaches that we can adopt to solve the problem. First, the method that we are probably most familiar with involves using the principles we learned in algebra, calculus, and other mathematics courses to manipulate the equations to determine the value(s) of the variable(s) of interest. For example, you have hopefully learned previously that when faced with the equation $2 + x = 5$, you simply subtract 2 from each side of the equation and establish that $x = 3$. This algebraic process that we have previously learned requires us to symbolically manipulate the equation until we arrive at the desired solution, hopefully.[1] This approach has the advantage of giving us an exact solution, but the disadvantage that it is limited to the set of problems where it is possible to obtain a solution through symbolic manipulations. The other approach involves determining an approximate solution, usually through an automatic iterative process on a computer. This approach is typically referred to as obtaining a numerical solution (although the careful reader may note that a better name is numerical approximate solution). The advantage of a numerical approach is that approximate solutions to a larger range of equations are possible, but the disadvantage is that the solutions are only approximate, the approach usually requires a computer, and the approach sometimes fails to find the desired solution.

It may be unusual to include a chapter on symbolic mathematics in a book that is focused on numerical methods, but for equations that can be solved by a symbolic approach, it is usually the preferred approach. Experienced engineers and mathematicians can usually determine relatively quickly if a set of mathematical equations is likely to be solvable using a symbolic approach. For novices, however, it is usually a good idea to try out a symbolic

---

1 An interesting historical example of mathematicians trying to use algebra and symbolic manipulation to solve the quintic equation can be found in "The Equation that Couldn't Be Solved" by Livio [1]. The roots of a quintic equation are typically easy to determine using a numerical process.

Chemical and Biomedical Engineering Calculations Using Python®, First Edition. Jeffrey J. Heys.
© 2017 John Wiley & Sons, Inc. Published 2017 by John Wiley & Sons, Inc.
Companion Website: www.wiley.com/go/heys/engineeringcalculations_python

approach, such as the one described in this chapter just to check-and-see if a symbolic solution is easily available. This chapter on symbolic computations also provides a good review of some Python principles that were covered previously, including the use of external libraries.

5.2 Symbolic Mathematics Packages

A large number of software packages have been developed for symbolic mathematics, and the capabilities of the various packages are not the same. As of 2014, Wikipedia listed over 30 different software packages for symbolic mathematics. The packages listed below are all commercial software, but they are among the most popular and site licenses are available on many university campuses.

Maple One of the oldest software packages for symbolic mathematics, and it was originally written at the University of Waterloo in the early 1980s. The name is a reference to Maple's Canadian heritage. While it was quite popular before 1995, its popularity declined due to a user interface that was difficult to use. The new user interface, introduced in 2005, is significantly better and similar to the other packages available for symbolic mathematics.

Mathcad One of the first mathematics software packages with a graphical user interface and support for SI units. The software is popular for producing reports and documentation that include mathematical calculations. The symbolic mathematics capabilities are sufficient for most purposes, but not as strong as some of the other packages listed here.

Mathematica Initially released in 1988, Mathematica was one of the first symbolic mathematics packages with a graphical user interface, which is referred to as the "front end". While the creation of custom algorithms remains difficult in Mathematica, it is still one of the most popular platforms for computational mathematics. Some of the functionality is available free-of-charge through the Wolfram Alpha website (https://www.wolframalpha.com/).

MATLAB's Symbolic Mathematics Toolbox MATLAB is primarily used for numerical computing, but the Symbolic Mathematics Toolbox provides some symbolic capabilities. Depending on the type of license, this is one of the most expensive options listed here.

One additional package that should be highlighted here is the Sage (previously SAGE, System for Algebra and Geometry Experimentation) mathematics software, which is free and licensed under a GNU General Public License. Of particular interest here is the fact that Sage uses the Python programming language, so individuals familiar with Python will have a more modest learning curve. Sage makes extensive use of Python libraries, including NumPy, SciPy, and SymPy, in order to avoid having to reimplement large amounts of existing code. While Sage is an excellent resource for mathematical computing, it is not covered in detail in this book because of the large size of the platform. The curious reader is encouraged to explore the Sage software and its features. The browser-based notebook interface available for Sage (SageMathCloud) may be of particular interest and was briefly discussed in Chapter 2.

The focus in this chapter is on the use of the SymPy library, which adds support for symbolic mathematics to Python [2]. SymPy is written entirely in Python and does not require any external libraries. SymPy is included with many distributions of Python that are focused on scientists or engineers including the Anaconda Python distribution and Pythonxy. Installation on Linux systems is also straightforward. Further information related to downloading and installing SymPy as well as comprehensive documentation is available on the SymPy website: www.sympy.org.

5.3   An Introduction to SymPy

The SymPy library is imported into any Python program that we write using the command: **import** sympy. As a result, all methods associated with the library are accessed using standard sympy.method() format. Alternatively, the entire library maybe imported as **from** sympy **import** * command, but use of this format is discouraged.

Once the SymPy library has been imported, the next step is to declare the symbolic variables or parameters that will be present in the equations that we plan to manipulate or solve symbolically. The sympy.symbols() class transforms a string that lists the variables or parameters into instances of the SymPy Symbol class. For example, the command: E, m, c = sympy.symbols('E m c') or E, m, c = sympy.symbols('E, m, c') converts the string "E m c" into three different symbolic variables that maybe used later to define mathematical expressions or equations or in future symbolic mathematics functions. It is strongly recommended that the symbol name (i.e., the variable on the left side of the "=" sign) be the same as the variable name in the string that is passed into the symbols() function.

Let us begin by demonstrating the SymPy library on a classic algebraic problem, factoring a quadratic polynomial, $a \cdot x^2 + b \cdot x + c = 0$, to determine its roots. As taught in a typical algebra course, the roots of a quadratic polynomial can be determined using the quadratic equation, $x = \frac{-b \pm \sqrt{b^2 - 4ac}}{2a}$. The derivation of the quadratic equation only requires straightforward algebraic manipulation of the quadratic polynomial to solve for x. The quadratic equation, that is, the roots of the quadratic polynomial, can also be derived using SymPy as is illustrated in the example below.

Determine the roots of the quadratic polynomial, $a \cdot x^2 + b \cdot x + c = 0$, using SymPy:

```
import sympy

a,b,c,x = sympy.symbols('a b c x')
expr = a*x**2 + b*x + c
print(sympy.solve(expr,'x'))
```

The quadratic polynomial is stored in the variable `expr`, and using the `solve()` function in the SymPy library allows for the determination of the roots of the polynomial. The output from the example code should be

```
[(-b+sqrt(-4*a*c + b**2))/(2*a), -(b+sqrt(-4*a*c +
b**2))/(2*a)],
```

which is a Python list containing the two roots of the polynomial.

If the values of a, b, and c are known, then the SymPy library may still be used. For example, the following modification to the code above solves for the roots of $3x^2 + 4x + 5 = 0$.

```
import sympy

x = sympy.symbols('x')
expr = 3*x**2 + 4*x + 5
print(sympy.solve(expr,'x'))
```

In this case, the output should be `[-2/3 - sqrt(11)*I/3, -2/3 + sqrt(11)*I/3]` where $I = \sqrt{-1}$. Note that the `sympy.solve()` function assumes that all terms in the equation have been moved to one side of the equal sign. If the equation in the following example had initially been written: $3x^2 + 5 = -4x$, it would have been necessary to first rearrange the terms so that they were all on the same side of the equal sign.

Another very helpful function in SymPy is the `subs()` method, which substitutes one expression for another. A simple example of this is replacing x in a polynomial with a specific value as illustrated below.

```
import sympy

x = sympy.symbols('x')
expr = 3*x**2 + 4*x + 5
print(expr.subs(x,2.0))
```

Here, 2.0 is substituted for x in the expression and the result is simplified to 25.000. The conversion of expressions into floating point numbers can use `evalf()`, and the desired precision can be passed into the function. For example, replacing the last line in the previous example with **print**(`expr.subs(x,2.0).evalf(16)`) evaluates the result after substitution to 16 digits of precision.

The `solve()` function was previously used for solving an algebraic equation. The syntax of this function is `solve(equations,variables)` where `equations` may be a single equation or a list of equations (a list in enclosed in square brackets, [item, item, item]). The number of variables listed, of course, must equal the number of equations. In the example below, the `solve()` function is used to solve a common problem in describing the behavior of gases.

The *van der Waals* equation of state is a common equation for relating the temperature (T), pressure (P), and specific volume ($\hat{V}$) of a nonideal gas. The equation may be written as

$$P = \frac{RT}{\hat{V} - b} - \frac{a}{\hat{V}^2},$$

where

$$a = \frac{27R^2 T_c^2}{64P_c},$$

$$b = \frac{RT_c}{8P_c},$$

$$R = 0.08206 \text{L} \cdot \text{atm}/(\text{mol} \cdot \text{K}).$$

and T_c and P_c are the critical temperature and pressure of the gas, respectively. Our goal is to calculate the specific volume of ammonia ($T_c = 405.5$ K and $P_c = 111.3$ atm) at $T = 420$ K and $P = 43.4$ atm. Before using SymPy to solve for the specific volume, we need to rewrite our equation(s) so all terms are on one side of the "=" sign. Thus, we will write the van der Waals equation as

$$0 = P - \frac{RT}{\hat{V} - b} + \frac{a}{\hat{V}^2}.$$

We are now ready to solve for the specific volume.

```
import sympy

R = 0.08206 # L atm /(mol K)
P = 43.4 # atm
T = 420.0 # K
Tc = 405.5 # K
Pc = 111.3 # atm

a = 27*(R**2 * Tc**2 / Pc)/64
b = R * Tc / (8 * Pc)

V = sympy.symbols('V')
f = P - R*T/(V-b) + a/(V**2)

print(sympy.solve(f,V))
```

The output from this example is

```
[0.70088, 0.06531 - 0.02985*I, 0.06531 + 0.02985*I]
```

The equation is cubic with respect to $\hat{V}$ so we should not be surprised by getting three solutions (i.e., three roots). In this case, it is simple to determine the correct solution as two of the solutions are complex and obviously not physical. We should also note that neither Python nor SymPy supports units, so the onus is on the user of the software to ensure that consistent and correct units are used in all calculations.

5.3.1 Multiple Equations

The `solve()` function is not limited to a single algebraic equation; it also supports multiple equations and unknowns. However, large systems of equations and unknowns are typically solved more efficiently using a numerical approach. The use of `solve()` for a relatively small and simple system of equations is demonstrated in the example below.

Use the `solve()` function to solve the following system of equations for x and y.

$$x^3 + y + 1 = 0,$$
$$y + 3x + 1 = 0.$$

It is important to note that both equations already have all terms on one side of the "=" sign.

```
import sympy

x, y = sympy.symbols('x y')
eq1 = x**3 + y + 1
eq2 = y + 3*x + 1
sol = sympy.solve([eq1, eq2], [x,y])
print(sol[0])
print(sol[1][0].evalf(), sol[1][1].evalf())
print(sol[2][0].evalf(), sol[2][1].evalf())
```

We should first observe that the solution is a list-of-lists: it is a list of three solutions and each solution is a list of the x, y-pair that satisfies the equations. In addition, note the use of `evalf()` to simplify the three solutions to this system of equations: $((0, -1), (-1.732, 4.196),$ *and* $(1.732, -6.196))$. We might be surprised that there are three different (x, y) pairs that satisfy this system of equations, but if we rearrange the equations slightly into

$$y_1 = -x^3 - 1$$
$$y_2 = -3x - 1$$

and plot the two curves using matplotlib.pylot, we can see that the two curves cross three times at the three solutions given previously (Figure 5.1).

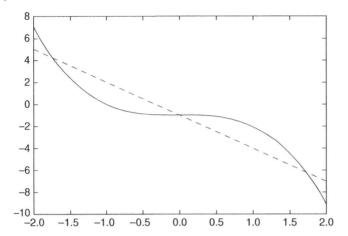

Figure 5.1 *x, y*-diagram of the two equations that were simultaneously solved in the multiple equations example.

5.4 Factoring and Expanding Functions

One of the more tedious and error-prone routine tasks in mathematics is expanding and factoring polynomial equations. SymPy can usually perform this task automatically. We can use SymPy to expand the function $f(x) = (x + 2)^3 + 3$ using the following code.

```
import sympy

x = sympy.symbols('x')
f = (x+2)**3 + 3
print(sympy.expand(f))
```

Running this short program gives us the expanded polynomial: $x^3 + 6x^2 + 12x + 11$. This expansion can also be performed by humans, but the error rate and time required are both high.

An even more useful feature of SymPy is its ability to factor polynomials. A student could spend hours trying to factor the polynomial $27x^3 + 135x^2 + 225x + 125$, but SymPy can factor it in a few seconds using the following code.

```
import sympy

x = sympy.symbols('x')
f = 27*x**3 + 135*x**2 + 225*x + 125
print(sympy.factor(f))
```

Running this code block prints the factorization, $(3x + 5)^3$ to the standard output. Using SymPy to expand and factor polynomials is simple enough that it can often be done straight from the Python prompt without the construction of a complete script.

5.4.1 Equilibrium Kinetics Example

In equilibrium kinetics, it is often necessary to expand and factor polynomial. Consider the water-gas shift reaction

$$CO + H_2O \rightleftharpoons CO_2 + H_2$$

proceeding to equilibrium at a temperature where the equilibrium coefficient is

$$K = \frac{y_{CO_2} y_{H_2}}{y_{CO} y_{H_2O}} = 1.4.$$

(This example is adapted from an example in Felder and Rousseau [3].) If the feed to this reactor is 2.0 mol of CO and 2.0 mol of H_2O and the extent that this reaction proceeds to the right (i.e., the extent of reaction) is ξ, then we can write the mole fraction of each species as

$$y_{CO} = 2 - \xi$$
$$y_{H_2O} = 2 - \xi$$
$$y_{CO_2} = \xi$$
$$y_{H_2} = \xi$$

and the equilibrium equation as

$$K = \frac{\xi \cdot \xi}{(2 - \xi)(2 - \xi)} = 1.4,$$

or

$$\xi^2 - 1.4(2 - \xi)^2 = 0.0.$$

The goal is to determine the extent of reaction, ξ, which is relatively straightforward but requires expanding out the polynomial. The following code illustrates the use of SymPy to either help us expand the polynomial or solve for ξ.

```
import sympy

xi = sympy.symbols('xi')
f = xi**2/((2-xi)*(2-xi)) - 1.4
g = xi**2 - 1.4*(2-xi)*(2-xi)
print(sympy.expand(g))
print(sympy.simplify(g))
print(sympy.factor(g))
print(sympy.solve(f))
```

The first three lines printed by the program all generate the simplified quadratic polynomial associated with the equation for ξ, specifically, they give $-0.4\xi^2 + 5.6\xi - 5.6$, which can be solved for ξ using the quadratic equation.

The last line prints out the solution, 1.084 or 12.92. Since we start with only 2.0 mols of CO and H_2O, the only physically possible solution is 1.084 mol.

5.4.2 Partial Fraction Decomposition

A challenging algebraic exercise that arises frequently in Process Controls (the area of Chemical and Biological Engineering that studies the automated control of process operations, including the use of sensors and valves) is the partial fraction decomposition of rational functions, which have the form:

$$f(x) = \frac{p(x)}{q(x)},$$

where $p(x)$ and $q(x)$ are polynomials with $q(x)$ being the higher degree polynomial. The objective of partial fraction decomposition is to decompose this rational function into one of the form:

$$f(x) = \frac{p(x)}{q(x)} = \frac{A}{x - k_1} + \frac{B}{x - k_2} + \cdots$$

by determining the roots of $q(x)$, which gives k_1, k_2, and so on, and the values for A, B, and so on. These rational functions are common when analyzing feedback control loops using Laplace transforms in process controls. The partial fraction decomposition is required for transforming the analysis from the Laplace domain into the real-world time domain.

The function `sympy.apart()` can automatically perform many partial fraction decompositions, and the use of this function is demonstrated in the example below, which decomposes the rational function $\frac{1}{s \cdot (s+0.5)}$:

```
import sympy

s = sympy.symbols('s')
f = 1/(s*(s+0.5))
print(sympy.apart(f))
```

The output from this example is

```
-2.0/(1.0*s + 0.5) + 2.0/s
```

and the interested reader can confirm that

$$\frac{1}{s \cdot (s + 0.5)} = \frac{-2}{s + 0.5} + \frac{2}{s}.$$

In this example, $k_1 = -0.5$, $k_2 = 0$, $A = -2$, and $B = 2$.

5.5 Derivatives and Integrals

It is also possible, and often very helpful, to use symbolic mathematics software when taking derivatives and integrals. Symbolic derivatives can be obtained using the `sympy.diff()` function. Let us begin by taking the derivative of a $\sin(x)$ function.

```
import sympy

x = sympy.symbols('x')
print(sympy.diff(sympy.sin(x),x))
```

The output from this code is $\cos(x)$, as expected. Passing in additional symbols (i.e., variables) into the `sympy.diff()` function causes additional derivatives to be taken. For example, if the last line in the previous code block is replaced with `print(sympy.diff(sympy.sin(x),x,x))`, the second derivative with respect to x is determined and the output is $-\sin(x)$. Alternatively, adding "y" to the symbols list and asking for the derivative with respect to "y", `print(sympy.diff(sympy.sin(x),x,y))`, gives the expected result of zero.

SymPy is especially helpful when taking derivatives of more complex functions because humans are more likely to make an error as the number of algebraic steps increases. The script below is used to take the derivative of $f(x) = x^2 \cdot \tan(x) + x \cdot \log(x)$:

```
import sympy

x = sympy.symbols('x')
f = x**2 * sympy.tan(x) + x * sympy.log(x)
print(sympy.diff(f,x))
```

and the output is the derivative:

```
x**2*(tan(x)**2 + 1) + 2*x*tan(x) + log(x) + 1
```

Note that in SymPy (and Python, in general), log (x) is the natural logarithm function.

5.5.1 Reaction Example

When designing a chemical reactor, we sometimes have a mathematical expression that relates the concentration of a species in the reactor to time. For example, assume that we know that

$$C_A = C_{A0} \exp(-k \cdot t),$$

where C_A is the concentration of species A, C_{A0} the initial concentration of A, k a constant, and t time. We would like to take the derivative of C_A with respect to time to determine the rate of the reaction. The following Python code will determine the derivative:

```
import sympy

Ca0, k, t = sympy.symbols('Ca0 k t')
```

```
Ca = Ca0 * sympy.exp(-k*t)
print(sympy.diff(Ca, t))
```

The output from the script is $-C_{A0} \cdot k \cdot \exp(-k \cdot t)$, which the observant reader will recognize can be simplified to $\frac{dC_A}{dt} = -k \cdot C_A$.

5.5.2 Symbolic Integration

Symbolic integration tends to be even more helpful than symbolic differentiation, probably because integration by hand is often more difficult than differentiation. The `sympy.integrate()` function is used for symbolic integration. The following code block demonstrates both the single and double integrations of a simple $\sin(x)$ function.

```
import sympy
```

```
x,y = sympy.symbols('x y')
print(sympy.integrate(sympy.sin(x),x))
print(sympy.integrate(sympy.sin(x),[x,0,1.0]))
print(sympy.integrate(sympy.sin(x),x,x))
print(sympy.integrate(sympy.sin(x),x,y))
```

Upon execution, the code outputs the expected result of $-\cos(x)$ for the first **print()** function call. The second **print()** function call is nearly identical to the first, but this time a list is given for the second argument. The list contains the symbolic variable to be integrated and the bounds on that variable. The result is a definite integral, and the output is 0.4597. The third **print()** function call results in $\sin(x)$ being integrated twice with respect to x, that is, this is a double integral, and outputs the expected result: $-\sin(x)$. What should we expect from the final print statement, which integrates $\sin(x)$ first against x and then against y? Well, the first integration will yield $-\cos(x)$, and the second integration will treat any function of x as a constant and, as a result, integration against y will give $-y \cdot \cos(x)$.

5.5.3 Reactor Sizing Example

As students learn in a reactor design course, the sizing of a batch reactor containing an irreversible second-order reaction $A \rightarrow B$ requires that we evaluate the integral:

$$\int_0^X \frac{1}{(1-X)^2} dX.$$

While this integral is still relatively simple, it never hurts to check our work with a symbolic mathematics program. The following script will evaluate this integral using SymPy.

```
import sympy

X = sympy.symbols('X')
t = sympy.integrate(1/((1-X)**2), (X, 0, X))
print(t)
print(sympy.simplify(t))
```

A couple interesting observations can be made from this block of code. First, notice the function interface for definite integrals: (variable, lower_bound, upper_bound), which, for the problem of interest here is (X, O, X). Second, the result of symbolic integration is initially (i.e., the first print statement): $-1 - \frac{1}{X-1}$, which is correct but not the simplest form possible. To obtain the more common, and simpler result, the sympy.simplify() function is used to get the standard result: $-\frac{X}{X-1}$.

5.6 Cryptography

The sympy.crypto library has a basic set of ciphers that allow for a gentle introduction to some classic cryptography. The one major constraint that the user of this library needs to be aware of is that the library only supports upper-case strings without spaces. Thus, before using any of the provided ciphers on a "secret message", it is necessary to convert the string into uppercase letters (using, e.g., the upper() function) and removing any spaces (using, for example, the replace(" ","") function where the first set of quotes encloses a single space and the second set of quotes is empty and without any spaces).

The first and simplest cipher is the shift cipher or Caesar cipher, after Julius Caeser, a purported user of the cipher. With this cipher, all the letters in the message are shifted n letters forward in the alphabet. If $n = 2$, for example, then "A" is replaced with "C" and "B" is replaced with "D". The key to reversing or deciphering the message is to reverse the shift, which only requires knowing n. The "key" for encoding and decoding shift cyphers is a single integer. The Python algorithm below demonstrates the use of the shift cipher.

```
import sympy.crypto.crypto as cipher

message = "secret code"
cleanMessage = message.upper().replace(" ","")
print(cleanMessage)
# Replace every letter with the next letter
print(cipher.encipher_shift(cleanMessage,1))
```

The output from the algorithm is:

```
SECRETCODE
TFDSFUDPEF
```

It is easy to confirm that every letter in the secret code was shifted to the next letter in the alphabet, and the key for this example is 1. Different keys can be specified in the call to the `shift()` function.

A slightly more complex cipher is the affine cipher, which requires two integers for the key that is used in encrypting and decrypting messages. The two integers, a and b, are used to map every letter, x, represented as an integer (0–25) to a new letter, y, also represented as an integer (0–25) using the linear function:

$$y = a \cdot x + b \ (\text{mod } 26).$$

If $a = 1$ and $b = 1$, then $y = x + 1$, which means that every letter is replaced with the same letter plus one, that is, this is identical to a shift cipher with a shift of 1. If $a = 2$ and $b = 2$ then $y = 2 \cdot x + 2$, and the letter "E", $x = 4$ is replaced with $y = 10$ or the letter "K". The use of the affine cipher is demonstrated with the script below.

```
import sympy.crypto.crypto as cipher

message = "secret message"
cleanMessage = message.upper().replace(" ","")
print(cleanMessage)
print(cipher.encipher_affine(cleanMessage,(2,2)))
```

The output of this script is

```
SECRETMESSAGE
MKGKKOAKMMCOK
```

confirming that "E" is replaced by "K" and the other letters are replaced as expected.

The final cipher explored here is the Vigenère cipher, named after Blaise de Vigenère. This cipher is similar to the shift cipher, but instead of shifting every letter by the same amount, a series of unique shift values are used instead. Further, instead of trying to remember a sequence of integers, the cipher's key is a string of letters, that is, a word, that is converted to a string of integers (0–25). Therefore, if the key is "CAB", it is converted to a series of integers: 2, 0, 1, and that series of integers is used repeatedly to shift the message. The use of this cipher is demonstrated in the Python script below.

```
import sympy.crypto.crypto as cipher

message = "secret message"
cleanMessage = message.upper().replace(" ","")
print(cleanMessage)
key = 'cab'# note 'cab' --> 2, 0, 1
result = cipher.encipher_vigenere(cleanMessage,key)
```

```
print(result)
print(cipher.decipher_vigenere(result,key))
```

The output from the script is:

SECRETMESSAGE
UEDTEUOETUAHG
SECRETMESSAGE

Some of the letters are not shifted due to the use of "a" in the key. The same key can be used to decipher or reverse the cipher and recover the original message.

None of the ciphers presented here is particularly secure or difficult for computers to decipher even without the key provided that the secret message is sufficiently long, but they demonstrate some of the basic concepts and they show the importance of lengthy keys or passwords, that is, do not use "cab" as a password.

Problems

5.1 You have been hired by the Mountain Chip company to analyze a new product they are bringing to market: Square chips! The Square chip has a similar shape to the traditional chip, but its exterior (perimeter) is square so it fits into a square box instead of a round can. The company expects to save millions on more efficient packaging and shipping.

The shape of the Square chip is described by the function: $f(x, y) = \frac{x^2}{1.0} - \frac{y^2}{2.0}$, where $-1.0 \leq x \leq 1.0$ and $-1.0 \leq y \leq 1.0$. The following Python algorithm plots the shape of the Square chip.

```
import sympy
x,y = sympy.symbols('x y')
z = x**2/1.0 - y**2/2.0
sympy.plotting.plot3d(z,(x,-1.0,1.0),(y,-1.0,1.0))
# Note that sympy.plotting.plot3d() calls
matplotlib
```

You have been hired as a consultant to answer two questions about the new Square chip by the Mountain Chip company.

a) The company believes that the Square chip is most likely to break where the curvature is greatest, and they would like you to determine the location(s) with the greatest curvature. Curvature is approximated by $C(x, y) = \left| \frac{\partial^2 f}{\partial x^2} \right| + \left| \frac{\partial^2 f}{\partial y^2} \right|$. Determine the location(s) and magnitude of the greatest curvature. The company made it very clear that they do not trust human calculations and they require an answer from computer using symbolic mathematics software.

b) In pursuit of ever greater packing efficiency, the company is concerned about the volume of air below the lowest chip in the stack of chips in the box. You need to determine the volume of air below the chip (i.e., the volume between the chip and a flat surface) by integrating the function describing the shape of the chip over the domain $-1.0 \leq x \leq 1.0$ and $-1.0 \leq y \leq 1.0$. It is probably necessary to modify the function so that the minimum value of the function $f(x, y)$ over the domain $-1.0 \leq x \leq 1.0$ and $-1.0 \leq y \leq 1.0$ is zero. This modified function will represent the chip resting on a flat surface (the flat surface is $f(x, y) = 0.0$). If you run the plotting function in the box above, you will see that the current function appears to possibly be less than zero for some values of $-1.0 \leq x \leq 1.0$ and $-1.0 \leq y \leq 1.0$.

5.2 You have been hired by a specialty chemical company that has been researching the physical properties of acetone (a common chemical for removing nail polish). The company observed that at an unknown temperature, acetone exhibited the same heat capacity as water. The company found the following quadratic equation for the heat capacity of acetone as a function of temperature [4]:

$$C_p = 26.63 + 0.183T - 45.86 \times 10^{-6} T^2 \ \text{J/(mol} \cdot \text{K)}$$

You have been hired to determine the temperature at which acetone has the same heat capacity as water (assume that water has a constant heat capacity of 75.6 J/(mol · K)) on a per mol basis. Finally, the company has two additional requirements: (1) you need to determine an equation for the change in heat capacity as a function of temperature, that is, $\frac{dC_p}{dT}$, and (2) plot the heat capacity as a function of temperature over the full range of temperatures where the heat capacity of acetone might equal the constant head capacity of water. The company is notoriously skeptical of people that perform mathematical analysis "by hand" and is requiring that you perform all calculations using symbolic mathematics software.

5.3 You have been hired by the recently reformed La Vie Claire cycling team (this is the team that Greg LeMond won the Tour de France with in 1986 – see the ESPN 30 for 30 documentary, "Slaying the Badger"). The new team manager knows a little physics and made an interesting observation while studying the standard fluid dynamics equation describing the drag force on a body moving through air (or any Newtonian fluid):

$$F_D = C_D A \frac{\rho V^2}{2.0}$$

where F_D, the drag force, is approximately equal to the force the rider is applying to the pedals (neglecting mechanical resistance) when riding on a flat surface. The team manager claimed that for a given force from the rider, the equation should have two solutions for the velocity, V, because it is quadratic. The manager further asserted that if riders could change their velocity somehow, they could shift their velocity to the other, faster solution to the equation without having to change the force on the pedals. You have been hired by the team owner to investigate this claim. Using symbolic mathematics software, show that there is only one positive velocity solution to the equation if F_D, the force, $C_D A$ (the drag coefficient multiplied by the rider's frontal or cross-sectional area), and ρ (the density of air, 1.0 kg/m^3) are all positive.

The second half of the owner's request is that you calculate the cyclist's velocity (in meters per second and miles per hour) using the following assumptions:

- Professional cyclists perform 10,000 kJ/day of work
- In a major race, cyclists ride 250,000 m/day
- Work is force times distance (i.e., force is work over distance)
- $C_D A$ for a cyclists alone on the road is 0.7–0.9 m^2

Finally, estimate the velocity of the same cyclist in a group where each cyclist can draft off the person in front of them, thus reducing $C_D A$ to 0.5–0.7m^2.

5.4 Have you ever looked at that little hole at the bottom of windows on commercial aircrafts and wondered, "Why is it there?" (answer: search 'holes in airplane windows' on http://www.slate.com). Commercial airplanes have three layers for each window: the inner layer to catch snot from sneezes, a middle layer that contains the tiny hole at the bottom, and an outer layer. It turns out that the hole in the middle layer is designed to be large enough to keep moisture from accumulating between the two outer layers while at the same time being small enough to prevent a total loss of cabin pressure in the event that the outer window layer completely fails.

You have been hired by a major airplane manufacturer to estimate the flow rate of air through the small hole in the event that the outer most layer of glass fails. The manufacturer needs to be sure that the airplane cabin pressurization system has the ability to prevent the total loss of cabin pressure.

The flow through the small, cylindrical hole should be estimated by assuming Poiseuille flow:

$$v = \frac{\Delta P}{4 \cdot \mu \cdot L}(R^2 - r^2),$$

where $\Delta P = 60$ kPa is the pressure difference between the inside and outside of the airplane, $\mu = 1.8 \times 10^{-5}$ kg/(m · s) the viscosity of air, $L = 1.0$ mm the length of the cylindrical hole, and $R = 0.5$ mm the radius of the hole. The velocity of air through the hole, v, is a function of the distance, r from the center of the hole. Hence, the velocity is maximum along the center of the hole and it decreases closer to the edges of the hole. At the edge of the hole, $r = R$ and the velocity is zero. The air near the edge is slowed down by friction with the glass of the window.

The first part of your contract with the airplane manufacturer is to use symbolic mathematics software to obtain an equation for the total flow through the hole, Q, by integrating the velocity across the cross section of the hole:

$$Q = 2\pi \int_0^R v \cdot r \cdot dr.$$

Then, determine the total flow rate in m^3/s and m^3/h using the properties given above.

5.5 In the field of Process Controls, it is sometimes necessary to perform what is called a partial fraction decomposition. Consider the equation

$$F(s) = \frac{s + 1}{s^2(s + 2)}.$$

The process of partial fraction decomposition requires that we determine the constants c_1, c_2, and c_3 such that the following equation is satisfied:

$$\frac{s + 1}{s^2(s + 2)} = \frac{c_1}{2s^2} + \frac{c_2}{4(s + 2)} + \frac{c_3}{4s}.$$

Fortunately, SymPy includes the function `sympy.apart()` that can usually take a partial fraction decomposition automatically. Write a script that determines the value of c_1, c_2 and c_3 using a partial fraction decomposition.

References

1 Livio, M. (2006) *The Equation that Couldn't be Solved: How Mathematical Genius Discovered the Language of Symmetry*, Simon & Schuster, New York, NY.

2 Team, S.D. (2014) *SymPy: Python library for symbolic mathematics*, http://www.sympy.org.

3 Felder, R.M. and Rousseau, R.W. (2005) *Elementary Principles of Chemical Processes*, John Wiley & Sons, Inc., Hoboken, NJ, 3rd edn.

4 Fogler, H.S. and Gurmen, M.H. (2015) *Elements of Chemical Reaction Engineering. Companion CD*, http://umich.edu/~elements/.

6

Linear Systems

A single, linear, algebraic equation is trivial to solve. In engineering, however, we are often faced with the more difficult challenge of solving for multiple unknowns (e.g., x_1, x_2, and x_3) that are related by multiple, linear algebraic equations. In the previous chapter on symbolic mathematics, we explored an approach, `sympy.solve()`, that gave an exact solution. This approach, however, is limited to problems with only a few equations and a few unknowns (typically <10). Our goal in this chapter is to learn methods that can handle thousands or even millions of unknowns.

If the equations are truly linear – the unknowns are not multiplied by each other or themselves, nor are there nonlinear terms within the equations, such as $\sin(x_1)$, then we can write the system of equations as

$$a_1 x_1 + a_2 x_2 + \cdots + a_n x_n = f_1$$
$$b_1 x_1 + b_2 x_2 + \cdots + b_n x_n = f_2$$
$$\vdots = \vdots$$
$$z_1 x_1 + z_2 x_2 + \cdots + z_n x_n = f_n,$$

where $a, b, \ldots, z$, and f each represent n-constants. The system has n-equations and n-unknowns. It is often simpler to write our this system of equations in the form of a matrix:

$$\begin{bmatrix} a_1 & a_2 & \cdots & a_n \\ b_1 & b_2 & \cdots & b_n \\ \vdots & & \ddots & \vdots \\ z_1 & z_2 & \cdots & z_n \end{bmatrix} \cdot \begin{bmatrix} x_1 \\ x_2 \\ \vdots \\ x_n \end{bmatrix} = \begin{bmatrix} f_1 \\ f_2 \\ \vdots \\ f_n \end{bmatrix}. \tag{6.1}$$

The matrix on the left-hand side and the right-hand-side vector both contain given constants, and the vector, x, in the middle contains the unknowns. This problem is often written $\mathbf{A} \cdot \mathbf{x} = \mathbf{f}$. Our goal in this chapter is to learn different approaches for solving for $\mathbf{x}$ whenever we have a system of linear equations.

Chemical and Biomedical Engineering Calculations Using Python®, First Edition. Jeffrey J. Heys.
© 2017 John Wiley & Sons, Inc. Published 2017 by John Wiley & Sons, Inc.
Companion Website: www.wiley.com/go/heys/engineeringcalculations_python

---

Note on Notation

Throughout this book, a bold lowercase variable (e.g., **x**) is used to represent a vector. A bold uppercase variable represents a matrix (e.g., **A**).

6.1 Example Problem

Distillation columns are used to separate mixtures of compounds based on differences in boiling points. The development of a mathematical model of a distillation column typically results in hundreds or thousands of linear and nonlinear equations. Let us explore a simplified mathematical model for a distillation column where the input is known: 30 kg/s of methane, 25 kg/s of ethane, and 10 kg/s of propane. The input mixture is separated into three outflow streams: a overhead stream that is rich in methane (90%) and does not contain any propane, a middle stream that is rich in ethane (50%) and a bottom stream that is rich in propane (70%). Propane is the least volatile of the three components in the distillation column and, hence, is the most likely to be separated into the bottom stream. Figure 6.1 contains additional information on the composition of the outflow streams – note that x is used for mass fractions (i.e., the fraction of a total stream that is a specific compound) and m is used for mass flow rates (in kg/s). Subscripts denote specific compounds – methane (M), ethane (E), and propane (P), or numerical subscripts represent different stream numbers so m_1 is the total mass flow rate of the entire stream 1.

Ideally, distillation columns are operated at steady state, and every kilogram of each compound that enters the column is matched by a kilogram of that same compound leaving the column. This must be true due to the conservation of mass. Using this principle, an equation that equates the mass flow rate of methane into the column to the mass flow rate of methane leaving the column can be written.

$$m_{in} = m_{out}$$
$$30 \text{ kg/s} = 0.9m_1 + 0.3m_2 + 0.1m_3. \tag{6.2}$$

The second equation utilizes the fact that the mass flow rate of methane in stream 1 must equal the total mass flow rate of that stream (m_1) multiplied by the fraction of the stream that is methane ($x_M = 0.9$). Since there are three outflow streams, the mass flow rate into the column must equal the combined mass flow rate from each of the three outflow streams – mass must be conserved!

Similarly, we can write mass conservation equations on ethane and propane also:

$$25 \text{ kg/s} = 0.1m_1 + 0.5m_2 + 0.2m_3, \tag{6.3}$$

$$10 \, \text{kg/s} = 0.0m_1 + 0.2m_2 + 0.7m_3. \tag{6.4}$$

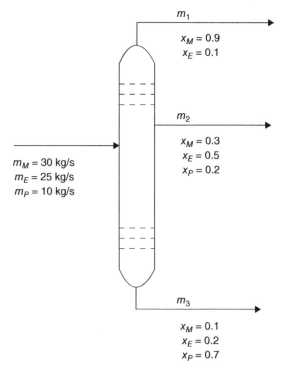

Figure 6.1 Diagram of a distillation column for the coarse separation of methane (M), ethane (E), and propane (P). The composition of the product streams is given, but the mass flow rate of each stream is unknown.

m_1
$x_M = 0.9$
$x_E = 0.1$

m_2
$x_M = 0.3$
$x_E = 0.5$
$x_P = 0.2$

$m_M = 30 \, \text{kg/s}$
$m_E = 25 \, \text{kg/s}$
$m_P = 10 \, \text{kg/s}$

m_3
$x_M = 0.1$
$x_E = 0.2$
$x_P = 0.7$

Note that there is no propane in stream 1, but we can still include stream 1 in the mass balance for propane (equation 6.4) by setting the fraction of the stream that is propane to 0.0. The result is a final system of three linear, algebraic equations with the same three unknowns. For simplicity, the system of equations (6.2–6.4) can be written in matrix form as

$$\begin{bmatrix} 0.9 & 0.3 & 0.1 \\ 0.1 & 0.5 & 0.2 \\ 0.0 & 0.2 & 0.7 \end{bmatrix} \cdot \begin{bmatrix} m_1 \\ m_2 \\ m_3 \end{bmatrix} = \begin{bmatrix} 30.0 \\ 25.0 \\ 10.0 \end{bmatrix}. \tag{6.5}$$

For review, let us use SymPy to solve for the flow rates of the three outflow streams.

```
import sympy

m1, m2, m3 = sympy.symbols('m1 m2 m3')
eq1 = 0.9*m1 + 0.3*m2 + 0.1*m3 - 30
eq2 = 0.1*m1 + 0.5*m2 + 0.2*m3 - 25
eq3 = 0.2*m2 + 0.7*m3 - 10
print(sympy.solve([eq1,eq2,eq3], [m1,m2,m3]))
```

SymPy symbolically solves this small system of equations and gives a solution of $m_1 = 17.9$ kg/s, $m_2 = 46.0$ kg/s, and $m_3 = 1.2$ kg/s. It is always a good idea to check a solution to make sure that the original equations are indeed satisfied.

Instead of symbolically solving this system, which is something that does not scale well to larger systems of equations, let us instead solve the system numerically. We will use the Numpy library (www.numpy.org) to build the required matrices and vectors for this process. The first step is to build the matrix $\mathbf{A}$, and the right-hand side, $\mathbf{f}$, and store them as numpy arrays.

```python
import numpy

A = numpy.array([[0.9,0.3,0.1],[0.1,0.5,0.2],
                 [0.0,0.2,0.7]])
f = numpy.array([30.0,25.0,10.0])
print(A)
print(f)
```

Notice that the entire $\mathbf{A}$ matrix is contained in a list, and the individual rows of the matrix are sublists (or nested lists) within the larger list.

Line Breaks in Python

Python supports two methods for breaking a long line of code up onto multiple lines (this is referred to as line wrapping or line continuation):

- parentheses, brackets, and braces can be used for automatic continuation because if the Python interpreter does not find a closing parenthesis on the same line as the opening parenthesis, it will automatically continue reading onto the next line as if the two lines are a single line. An example is

```python
a = [1, 2, 3, 4,
     5, 6]
```

- placing a backslash, "\", at the end of a line causes the Python interpreter to continue reading the next line as if it were on the same line. An example is

```python
a = "Hello" + \
    "World"
```

To solve this system of equations numerically, we need to import an additional library that contains common linear algebra functions. There are a number of linear algebra libraries available for Python, but one easy to use library is distributed with numpy and can be imported using **import** numpy.linalg. Note that this library is not automatically loaded when we **import** numpy and must be imported separately. Most people tire of typing numpy.linalg.function() over-and-over, so it is common to import this library using a shorter name such as "nl". This is possible using the command **import** numpy.linalg as nl. The following code uses the nl.solve() function to solve the matrix problem and determine the unknown flow rates.

```
import numpy
import numpy.linalg as nl

A = numpy.array([[0.9,0.3,0.1], [0.1,0.5,0.2],
                 [0.0,0.2,0.7]])
f = numpy.array([30.0,25.0,10.0])
x = nl.solve(A,f)
print(x)
```

The output is a vector containing the three unknown flow rates: [17.88 45.96 1.15].

The nl.solve() function computes the "exact" solution of a well-determined linear matrix equation, $\mathbf{A} \cdot \mathbf{x} = \mathbf{f}$. The term "exact" is in quotes because the solution is only "exact" up to computer round-off error. In other words, the solution will typically have 8–12 digits of accuracy depending on the condition of the matrix, type of computer, and other factors. Methods that compute an "exact" solution to a linear matrix equation are called *direct* methods. In the next section, we will examine some of the principles behind direct methods and discuss their scaling. Direct methods are a good choice for systems of 2 to 10,000 equations (although this range changes with available computational power). The computational algorithm used by the numpy.linalg.solve() function is a common LAPACK routine that is written in FORTRAN. Because the underlying algorithm is written in a compiled language instead of Python, it is more computationally efficient and scalable.

6.2 A Direct Solution Method

The goal of this section is to briefly examine a simple algorithm for directly calculating the "exact" solution to a matrix problem. Even though the algorithm presented here is significantly simpler than the more complex algorithms contained in LAPACK and used by numpy, it will still be the most complex Python code written up to this point in this book. The direct solver will actually be split into two different functions – a Gaussian elimination function and a backward substitution function. It is recommended that the reader create an empty Python file (a suggested filename is bobcatSolver.py) that contains both of the functions. This file or *module* can then be imported into other Python codes and the functions within it called using **import** bobcatSolve and then bobcatSolve.functionname().

The first function that will be written implements Gaussian elimination. To illustrate the process of Gaussian elimination, recall the example matrix problem from the distillation example that was derived previously.

$$\begin{bmatrix} 0.9 & 0.3 & 0.1 \\ 0.1 & 0.5 & 0.2 \\ 0.0 & 0.2 & 0.7 \end{bmatrix} \cdot \begin{bmatrix} m_1 \\ m_2 \\ m_3 \end{bmatrix} = \begin{bmatrix} 30.0 \\ 25.0 \\ 10.0 \end{bmatrix}. \tag{6.6}$$

In the linear equation system, each row of $\mathbf{A}$ and $\mathbf{f}$ represents an equation or *equality*. It is perfectly mathematically reasonable to multiply an entire equation by a constant or add/subtract one equation from another without changing the solution. Our goal is to multiply equations by a constant and then add or subtract equations from each other so that the lower triangular part of the matrix is zero – in other words, we want a matrix that is all zeros below the main diagonal. For the matrix in equation 6.6, the main diagonal has the values: 0.9, 0.5, and 0.7. Let us start by eliminating the value in the first column that is directly below the main diagonal – the value is currently 0.1. Observe that if we multiple the first equation (row 1) by $\frac{0.1}{0.9}$ and then subtracting the resulting equation from the second row, we will eliminate the 0.1 value in the first column and directly below the main diagonal. Specifically, if R_1 and R_2 represent rows 1 and 2, respectively, then replacing R_2 with $R_2 - R_1 \cdot \frac{0.1}{0.9}$ gives

$$\begin{bmatrix} 0.9 & 0.3 & 0.1 \\ 0.0 & 0.4667 & 0.1888 \\ 0.0 & 0.2 & 0.7 \end{bmatrix} \cdot \begin{bmatrix} m_1 \\ m_2 \\ m_3 \end{bmatrix} = \begin{bmatrix} 30.0 \\ 21.667 \\ 10.0 \end{bmatrix}. \tag{6.7}$$

Notice that row 1 (representing equation 6.6) did not change at all; the only change was to row 2. This process can now be repeated for all nonzero terms below the main diagonal – a process called Gaussian elimination. The second step would normally be to eliminate the term in column 1, row 3, but that term is already 0.0 in this example so no action is required. The third and final step for this example is to eliminate the term in column 2, row 3, which currently has a value of 0.2. This term is eliminated by multiplying row 2 by $\frac{0.2}{0.4667}$ and replacing row 3 by row 3 minus row 2 times this value (i.e., replacing R_3 with $R_3 - \frac{0.2}{0.4667} \cdot R_2$). Note that row 1 is not used in this elimination step because its use would introduce a nonzero value into column 1, row 3 of the matrix – a term that was just eliminated in step 2. In Gaussian elimination, the nonzero terms below the main diagonal are eliminated using same row as the column where terms are being eliminated. For example, the nonzero terms below the main diagonal in column 2 are eliminated using row 2.

A similar process is called LU-decomposition, which refers to the decomposition of a matrix into a lower triangular matrix (L) and an upper triangular matrix (U). The terms "Gaussian elimination" and "LU-decomposition" are frequently used interchangeably, although they are not exactly the same algorithm. For the current example problem, the result after Gaussian elimination is

$$
\begin{bmatrix} 0.9 & 0.3 & 0.1 \\ 0.0 & 0.4667 & 0.1888 \\ 0.0 & 0.0 & 0.619 \end{bmatrix} \cdot \begin{bmatrix} m_1 \\ m_2 \\ m_3 \end{bmatrix} = \begin{bmatrix} 30.0 \\ 21.667 \\ 0.714 \end{bmatrix}. \tag{6.8}
$$

Examination of the linear matrix system 6.8 shows that solving for the vector of unknowns, $[m_1, m_2, m_3]$, is now relatively trivial. Starting with the last equation, which is now $0.619m_3 = 0.714$, we can easily solve for $m_3 = 1.15$. Once m_3 is determined, it becomes trivial to solve for $m_2 = (21.667 - 0.188m_3)/0.4667$. This process of solving for the final solution after Gaussian elimination is referred to as backward substitution.

Python code for a simple Gaussian elimination function, called bobcatLU, is given below.

```python
import numpy

def bobcatLU(A,f):
    n = f.size
    # check for compatible matrix and rhs sizes
    if (A.shape[0] != n or A.shape[1] != n):
        print("Error! Incompatible input sizes.")
        return f
    # Loop through the columns of the matrix
    for i in range(0,n-1):
    # Loop through rows below diagonal for each column
        for j in range(i+1,n):
            if A[i,i] == 0:
                print("Error: Zero on diagonal!")
                print("Need algorithm with pivoting")
                return f
            m = A[j,i]/A[i,i]
            A[j,:] = A[j,:] - m*A[i,:]
            f[j] = f[j] - m*f[i]
    return A,f
```

The function receives a matrix, **A**, and a right-hand side, **f**, as inputs. The first few lines of the code check to ensure that the matrix and the right-hand side have a compatible size. Next, a loop through the columns of the matrix (with the exception of the last column that does not have any terms below the main diagonal) is initiated. For each column, *i*, there is a second loop (*j*) through the rows below the main diagonal. For every terms below the main diagonal, the term is eliminated using row *i* (i.e., equation *i*) multiplied by the appropriate multiplier, *m*. After Gaussian elimination is complete, the modified matrix **A** and the right-hand side, **f**, are returned.

Multiple Return Variables

The bobcatLU function returned multiple variables with: `return` A, f, which can also be written as `return` (A, f). In either case, a Python container, specifically a tuple, is returned that contains both variables. When a function returns multiple arguments, it is recommended that enough variables are defined to hold the individual return arguments. When calling bobcatLU, for example, use:

```
M,n = bobcatLU(A,f)
```

where A, f are passed *into* the function, and M, n are the variables returned *from* the function.

Backward substitution is an algorithm of similar complexity and is given below.

```
import numpy

def bobcatBS(A,f):
    n = f.size
    # Check for compatible matrix and rhs sizes
    if (A.shape[0] != n or A.shape[1] != n):
        print("Error! Incompatible input sizes.")
        return f
    # initialize the solution vector, x, to zero
    x = numpy.zeros((n,1))
    # solve for last entry first
    x[n-1] = f[n-1]/A[n-1,n-1]
    # loop from the end to the beginning
    for i in range(n-2,-1,-1):
        sum = 0
        # for known x values, sum and move to rhs
        for j in range(i+1,n):
            sum = sum + A[i,j]*x[j]
        x[i] = (f[i] - sum)/A[i,i]
    return x
```

The backward substitution algorithm begins by checking the dimensions of the input parameters and initializing a vector, $\mathbf{x}$, that will ultimately hold the solution. Then, starting with the last row in the linear matrix system, the algorithm calculates the corresponding value for the $\mathbf{x}$-vector. The algorithm proceeds from the last row to the first row before completing.

It is simplest to combine the Gaussian elimination and backward substitution algorithms into a single file. Note that only a single **import** numpy command is required at the start of the file. The resulting file is called a *module* in Python programming, and it can be imported and used with other code. This is a very simple and efficient mechanism for recycling code. As an example, if the bobcatLU() and bobcatBS() algorithms are saved in a file called bobcatSolve.py, then the algorithms can be used to solve the previous distillation column example in a straightforward manner as illustrated in the example below.

6.2.1 Distillation Example

Use the bobcatLU() and bobcatBS() functions to solve the distillation column example problem.

```python
import numpy as np
import bobcatSolve as bS

A = np.array([[0.9,0.3,0.1], [0.1,0.5,0.2],
              [0,0.2,0.7]])
f = np.array([30.0,25.0,10.0])

A,f = bS.bobcatLU(A,f)
x = bS.bobcatBS(A,f)

print(x)
```

The solution should be the same as obtained using the numpy.linalg.solve() function: [17.9, 46.0, 1.2].

6.2.2 Blood Flow Network Example

A large number of mathematical models of blood flow have been developed. Some of these models are highly complex and account for the flexibility of the blood vessel walls, the impacts of blood cells, and the effects of reflected pressure waves on flow. Other models of blood flow are less accurate because a large number of assumptions have been made to simplify the mathematical model. The simplest model of blood flow assumes that the flow is steady (not pulsatile), the vessel walls are rigid, and the blood vessels are straight cylinders. Under these assumptions, the flow can be approximated using the Poiseuille flow solution [1], which states that the flow rate through the vessel is proportional to the pressure decrease, given by the equation:

$$\Delta P = \left(\frac{128 \cdot \mu \cdot L}{\pi \cdot d^4} \right) \cdot Q, \tag{6.9}$$

where ΔP is the pressure decrease, μ the viscosity of blood (4 dyn $\cdot$ s/cm^2), L the length of the vessel, and d the diameter of the vessel.

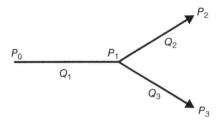

Figure 6.2 Diagram of a simple network of three blood vessels: the flow is into vessel 1 (left) at pressure P_0, and at the end of vessel 1, there is a bifurcation or branch and the flow is divided between vessels 2 (upper right) and 3 (lower right). The pressures, P_i, at the ends of each vessel as well as the flow rate, Q_i, in each vessel are potential unknowns that need to be determined.

Our goal is to model flow through the three vessels shown in Figure 6.2. Vessel 1, on the left, represents the femoral artery, and it has an unknown flow rate, Q_1, and pressure drop, $\Delta P = P_0 - P_1$. Vessel 1 then branches (or bifurcates) into two smaller vessels: the upper vessel (vessel 2) has an unknown flow rate Q_2 and pressure drop, $\Delta P = P_1 - P_2$, and the lower vessel (vessel 3) has an unknown flow rate Q_3 and pressure drop, $\Delta P = P_1 - P_2$.

The blood pressure at the entry to the femoral artery, $P_0 = 5000 \text{ dyn/cm}^2$, is relative to the pressure at the outflow, $P_2 = P_3 = 0 \text{ dyn/cm}^2$. Table 6.1 summarizes the available geometric data on the three vessels. Starting with the femoral artery (vessel 1), the value of most variables in the Poiseuille flow equation (6.9) can be specified, and the equation can be simplified to

$$\frac{128 \cdot \mu \cdot L_1}{\pi \cdot d_1^4} \cdot Q_1 = 637 \cdot Q_1 = 5000 - P_1. \tag{6.10}$$

Similarly, the Poiseuille flow equation for the upper and lower branches can be simplified to

$$\frac{128 \cdot \mu \cdot L_2}{\pi \cdot d_2^4} \cdot Q_2 = 2121 \cdot Q_2 = P_1 - 0, \tag{6.11}$$

$$\frac{128 \cdot \mu \cdot L_3}{\pi \cdot d_3^4} \cdot Q_3 = 2121 \cdot Q_3 = P_1 - 0. \tag{6.12}$$

Table 6.1 Properties of the femoral artery and upper and lower branches.

Vessel	Diameter (cm)	Length (cm)
1 (femoral artery)	0.4	10
2 (upper branch)	0.28	8
3 (lower branch)	0.28	8

Examining the three equations (6.10–6.12), we should note that there are four unknowns: Q_1, Q_2, Q_3, and P_1, so we need one additional equation to have a solvable system of linear equations. The final equation recognizes that the flow through the femoral artery must equal the sum of the flow through the two branches, that is, $Q_1 = Q_2 + Q_3$, using the assumption that blood in this system has constant density.

The system of equations given above for determining blood flow in the three vessels can be written as a matrix, a vector of unknowns, and a right-hand side. It is *critical* that the equations above be rearranged slightly by moving all terms with unknowns to one side of the equal sign. Hence, the P_1 term in the first three equations must be moved to the left side, and the Q_2 and Q_3 terms in the fourth equation must be moved to the left side.

$$
\begin{bmatrix}
637 & 0.0 & 0.0 & 1.0 \\
0.0 & 2121 & 0.0 & -1.0 \\
0.0 & 0.0 & 2121 & -1.0 \\
1.0 & -1.0 & -1.0 & 0.0
\end{bmatrix}
\cdot
\begin{bmatrix}
Q_1 \\ Q_2 \\ Q_3 \\ P_1
\end{bmatrix}
=
\begin{bmatrix}
5000 \\ 0.0 \\ 0.0 \\ 0.0
\end{bmatrix}. \tag{6.13}
$$

The Python code below solves the linear system of equations using both the "bobcatLU()" function and the numpy.linalg.solve() function. The reader should note that even though the original matrix has a term on the main diagonal that is equal to zero, the term becomes nonzero during the Gaussian elimination process and the "bobcatLU()" function does not give an error.

```python
import numpy
import bobcatSolve as bS
import numpy.linalg as nl

P0 = 5000.0 # dynes/cm^2
L1 = 10.0 # cm
L2 = L3 = 8.0 # cm
d1 = 0.4 # cm
d2 = d3 = 0.28 # cm
mu = 0.04 # dyn*s/cm^2
R1 = (128 * mu * L1) / (numpy.pi * d1**4)
R2 = (128 * mu * L2) / (numpy.pi * d2**4)
R3 = (128 * mu * L3) / (numpy.pi * d3**4)
# Unknowns: Q1, Q2, Q3, P1 with Q in mL/s

A = numpy.zeros((4,4), dtype=numpy.float)
A[0,0] = R1
A[0,3] = 1.0
A[1,1] = R2
A[1,3] = -1.0
A[2,2] = R3
A[2,3] = -1.0
```

```
A[3,0] = 1.0
A[3,1] = A[3,2] = -1.0
f = numpy.array([P0, 0.0, 0.0, 0.0])
print(nl.solve(A,f))
A,f = bS.bobcatLU(A,f)
x = bS.bobcatBS(A,f)
print(x)
```

The solution after solving for the unknowns is $Q_1 = 2.9 \, \text{mL/s}$, $Q_2 = Q_3 = 1.5 \, \text{mL/s}$ and $P_1 = 3100 \, \text{dyn} \cdot \text{s/cm}^2$. These values are consistent with experimental measurements [2].

6.2.3 Computational Cost

Gaussian elimination and backward substitution are much more computationally efficient than symbolic computing, but the computational scalability is still not optimal. A very rough approximation of the computational cost can be made by examining the Gaussian elimination algorithm. The elimination of all nonzero terms below the main diagonal requires looping through on the order of n-columns and n-rows. For each entry, there are approximately n-multiplications, so the total computational cost is on the order of n^3 operations. It is common to use the shorthand notation: $O(n^3)$ for something that is on the order of n^3.

To test this estimate of computational cost, the code below was used to measure the computational time for solving an increasing number of linear equations. The code uses the Python time library to determine the solve time by calculating the difference between the start time and stop time of a calculation. The problem is based on a dense matrix of random numbers, and a random right-hand-side vector, and the smallest problem is 100 equations and the largest is 6400 equations.

```
import numpy
import numpy.linalg as nl
import time
import pylab

mag = 4
cputime = numpy.zeros((mag,1))
cpusize = numpy.zeros((mag,1))
n=100
for i in range(mag):
    print(n)
    A = numpy.random.random((n,n))
    b = numpy.random.random((n,1))
    start = time.clock()
```

```
      x = nl.solve(A,b)
      stop = time.clock()
      cputime[i]=stop-start
      cpusize[i]=n
      n = n*4
pylab.loglog(cpusize,cputime)
pylab.xlabel('Number of Equations')
pylab.ylabel('CPU time (sec.)')
```

The CPU time measurements versus number of equations are summarized in Figure 6.3, which was obtained on a Dell laptop with a Core i5 CPU. The smallest problem size (100 equations) required only 0.0006 s. Assuming that the scaling of the algorithm is n^3, increasing the problem size by a factor of 4 should increase the computational time by a factor of $4^3 = 64$. The observed CPU time increase is closer to a factor of 36 when going from 100 to 400 equations, but the observed increase is exactly a factor of 64 when going from 1600 to 6400 equations. For this particular computer, 6400 equations required about 1 min, which is why direct methods are rarely used for problems larger than approximately 10,000 equations (unless the matrix is sparse, that is, contains mostly zeros).

The n^3 scaling of direct methods motivates the development of alternative approaches that give up the goal of obtaining an "exact" solution in exchange for improved scaling. In the final section of this chapter, iterative methods that

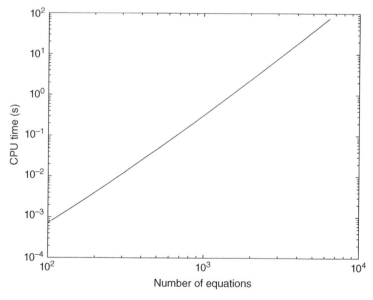

Figure 6.3 The CPU time required to solve a dense system of linear equations using numpy.linalg.solve(). If n is the number of equations, the CPU time scales with n^3.

can, in some cases, improve the scaling of CPU time relative to problem size will be briefly examined.

6.3   Iterative Solution Methods

The basic motivation behind iterative methods is the observation that the computational cost of multiplying a matrix and vector is on the order of n^2-multiplications for a dense matrix and on the order of n-multiplications for a matrix that is sparse (i.e., a matrix that contains mostly zeros). If we have a guess at the solution vector, which we can refer to as $\mathbf{x}_0$, then it is computationally inexpensive to calculate what is referred to as the residual: $\mathbf{res} = \mathbf{f} - \mathbf{A} \cdot \mathbf{x}_0$. Notice that the residual is a measure of how close our guess, $\mathbf{x}_0$, is to satisfying the original matrix problem, $\mathbf{A} \cdot \mathbf{x} = \mathbf{f}$. If the values in the residual vector are "small", the guess is close to the solution. In order to define "small" in more specific or quantitative terms, we need to briefly discuss norms.

6.3.1   Vector Norms

A norm is a single number that reflects the size of a vector. The most commonly used norm is the L^2-norm or Euclidean norm and it is calculated as

$$|\mathbf{x}| = \sqrt{\sum_{k=1}^{n} (x_k)^2}.$$

The L^2-norm of a Numpy vector, $\mathbf{x}$, is calculated using `numpy.norm(x,2)`, where the "2" denotes the L^2-norm. One other norm of notable relevance to the calculations of interest in this book is the infinity norm, which is calculated by finding the term in a vector with the largest absolute value. The infinity norm of a Numpy vector is calculated using `numpy.norm(x,numpy.inf)`.

6.3.2   Jacobi Iteration

To illustrate our first iterative method, let us return to the system of equations that represent mass balances around a distillation column.

$$30 \text{ kg/s} = 0.9m_1 + 0.3m_2 + 0.1m_3,$$
$$25 \text{ kg/s} = 0.1m_1 + 0.5m_2 + 0.2m_3,$$
$$10 \text{ kg/s} = 0.0m_1 + 0.2m_2 + 0.7m_3.$$

One approach to determine values for the unknowns: $m_1 - m_3$, would be to make an initial guess, for example, that $m_1 = 20$ kg/s, $m_2 = 20$ kg/s, and $m_3 = 20$ kg/s. This is not the most reasonable of guesses since total mass is not conserved (i.e., more mass is flowing into the column than out based on our crude guess), but the goal is to illustrate that our guess does not necessarily need to be *really* close to the actual solution. Now, let us solve the first equation for m_1, using our guess for the values of m_2 and m_3. It is trivial to calculate a new guess for $m_{1,new} = (30 - 0.3 \cdot 20 - 0.1 \cdot 20)/0.9 = 24.44$. Repeating this process and solving for a new guess for m_2 using the second equation and the old guess for both m_1 and m_3 results in $m_{2,new} = 38$ and, finally, $m_{3,new} = 8.6$ using the third equation. Notice that the new guess is indeed closer to the solution determined previously in this chapter than our initial guess of 20 kg/s for every stream. If we repeated this process a "few" more times, always using our improved guess, we might converge toward the "exact" solution.

The Python script below will help us to perform these calculations quickly and automatically (and we will learn some new Python programming practices as well!).

```python
import numpy
import numpy.linalg as nl

def jacobi(A,f,x,maxIter = 100, tol = 1.0e-4):
    #   inputs:
    #   A is a nxn matrix
    #   f is a right-hand-side vector of length n
    #   x is initial guess at the solution to A x = f
    #   maxIter (optional) is maximum iterations
    #   tol (optional) is desired accuracy in terms
    #   of the L2-norm of the residual (= f - Ax)
    n = f.size
    # Begin by checking for compatible sizes
    if (A.shape[0] != n or A.shape[1] != n):
        print("Error! Incompatible sizes.")
        return f
    # Loop to iterate until we converge to solution
    # or we reach the maximum number of iterations
    xnew = numpy.copy(x)
    for iter in range(maxIter):
        # calculate residual
        res = f - numpy.dot(A,x)
        # check L2-norm for convergence
        if (nl.norm(res,2) < tol):
            print('Converged after ', iter,
                ' iterations')
            return x
        # start of Jacobi iteration
        for i in range(n):
            sum=0.0
            for j in range(n):
                if(i != j):
                    sum += A[i,j]*x[j]
            xnew[i] = (f[i] - sum)/A[i,i]
```

```
        x = numpy.copy(xnew)
    print('Failed to converge after ', iter,
        ' iterations')
    return x

A = numpy.array([[0.9, 0.3, 0.1], [0.1,0.5,0.2],
                [0.0,0.2,0.7]])
f = numpy.array([30.0,25.0,10.0])
x = numpy.array([20.0, 20.0, 20.0])
sol = jacobi(A,f,x)
print(sol)
```

The code begins with the definition of a function, called jacobi, but, it is important to emphasize that when we execute or run this code, execution actually begins with the line that constructs the matrix **A**. The function definition is read by Python and stored for later use, but the function is not executed until it is called in the second to the last line of the script. The function definition must appear before the function is first called because, otherwise, Python will return an error stating that the function has not been defined when it is first called. The function itself requires that at least three variables be passed into the function, a matrix, a right-hand-side vector, and a vector containing a guess at the solution. However, the function allows two additional, optional arguments to be passed when it is called. The first optional argument is the maximum number of Jacobi iterations, and the default value is set to 1000 if another value is not passed into the function. The second optional argument is the desired tolerance, that is, the maximum L^2-norm of the residual vector for acceptable convergence. The jacobi function iterates until either the maximum number of iterations is reached or the desired tolerance is achieved, whichever is reached first. A helpful comment at the top of the function reminds the user of the input variable requirements.

The lines of Python code for the first half of the function are largely comments or code that we have used before. An iteration loop is initiated to run for at most the maximum number of iterations allowed and then the residual vector and its L^2-norm are calculated. Before performing the calculations associated with the Jacobi iteration, the L^2-norm of the residual is always calculated to test for convergence. If the norm is less than the desired tolerance, the current guess is returned and the function execution ends. If the norm is not less than the tolerance, the coefficients in **A** are multiplied by the current guess at the solution (except for the coefficient on the diagonal associated with the unknown we are determining) and these are subtracted from the right-hand side and divided by the coefficient along the diagonal. The reader is encouraged to revisit the process described above for solving for one unknown for each of the mass balance equations and to observe the connection to the Jacobi iteration in the Python script. If the desired solution tolerance is not achieved after the maximum number of iterations has been reached, the function prints an error message and simply returns the (incorrect) vector **x** after the final iteration.

Testing the Jacobi iterative method on the distillation column mass balances results in 14 iterations being required to achieve the default tolerance for the L^2-norm of the residual. It is interesting to test the method with different initial guesses for the solution. For example, if our initial guess had been

```python
x = numpy.array([10.0, 10.0, 10.0])
```

the algorithm would have required 16 iterations to achieve a solution satisfying the same tolerance.

Robustnesses of Iterative Methods

Test the Jacobi iterative method on the following matrix and right-hand side:

$$\begin{bmatrix} 1.0 & -1.0 & 2.0 & -1.0 \\ 2.0 & -2.0 & 3.0 & -3.0 \\ 1.0 & 1.0 & 1.0 & 0 \\ 1.0 & -1.0 & 4.0 & 3.0 \end{bmatrix} \cdot \begin{bmatrix} x_1 \\ x_2 \\ x_3 \\ x_4 \end{bmatrix} = \begin{bmatrix} -8.0 \\ -20.0 \\ -2.0 \\ 4.0 \end{bmatrix}. \tag{6.14}$$

The desired solution is $[-7.0, 3.0, 2.0, 2.0]$, but for almost any initial guess (other than the exact solution), the Jacobi iteration fails to converge. The cause of this failure is described below in the section on Convergence of Iterative Methods. The simplest solution to this failure is to use a LU-decomposition or some other direct solver.

6.3.3 Gauss–Seidel Iteration

The Jacobi iteration calculates a new guess for the vector **x** based *only* on the previous guess. It is therefore possible to compute each entry in the new guess vector simultaneously. An obvious alternative to this approach is to calculate a new value for the first entry in the unknown vector **x** but then use this new value for calculating the second entry in the vector **x**. Continuing in this manner, each new value in **x** is always calculated using the most recent information available. For the first iteration of the distillation column example, the calculation of $m_1 = 24.4$ would be identical to the Jacobi iteration, but the calculation of m_2 using the new value for m_1 would result in $m_2 = 37.11$ instead of $m_2 = 38$.

The implementation of the Gauss–Seidel iteration is nearly identical to the implementation of the Jacobi iteration, except the vector xnew is no longer required since all calculations involve only the most recent information that is already stored in x. The Python script that implements the Gauss–Seidel iteration for the distillation column mass balances is given below.

```python
import numpy
import numpy.linalg as nl

def gaussSeidel(A,f,x,maxIter = 100, tol = 1.0e-4):
    # inputs:
    # A is a nxn matrix
    # f is a right-hand-side vector of length n
    # x is initial guess at the solution to A x = f
    # maxIter (optional) is maximum iterations
    # tol (optional) is the desired accuracy in terms
    # of the L2-norm of the residual (= f - Ax)
    n = f.size
    # Begin by checking for compatible sizes
    if (A.shape[0] != n or A.shape[1] != n):
        print("Error! Incompatible sizes.")
        return f
    # Loop to iterate until we converge to solution
    # or we reach the maximum number of iterations
    for iter in range(maxIter):
        # calculate residual
        res = f - numpy.dot(A,x).flatten()
        # check L2-norm for convergence
        if (nl.norm(res,2) < tol):
            print('Converged after ', iter,
                   ' iterations')
            return x
        # start of Gauss-Seidel iteration
        for i in range(n):
            sum=0.0
            for j in range(n):
                if(i != j):
                    sum += A[i,j]*x[j]
            x[i] = (f[i] - sum)/A[i,i]
    print('Failed to converge after ', iter,
           ' iterations')
    return x

A = numpy.array([[0.9, 0.3, 0.1], [0.1, 0.5, 0.2],
                 [0.0, 0.2, 0.7]])
f = numpy.array([30.0,25.0,10.0])
x = numpy.array([20.0, 20.0, 20.0])
sol = gaussSeidel(A,f,x)
print(sol)
```

Applying the Gauss–Seidel iteration to the test problem results in eight iterations being required for convergence to the approximate solution with the same default tolerance used previously (recall that 10 Jacobi iterations were required). The Gauss–Seidel iterative method typically converges with significantly fewer iterations and over a greater range of initial guesses than the Jacobi iterative method. The only reason to use Jacobi iterations instead of Gauss–Seidel iterations is that calculations involved in the Jacobi iteration may be executed in parallel, which may result in shorter computational times on some computer architectures even with a larger total number of iterations.

6.3.4 Relaxation Methods

Iterative methods are based on the idea of improving our guess for the solution each iteration. For Gauss–Seidel, the improved guess is obtained through the calculation: `x[i] = (f[i] - sum)/A[i,i]`. Informally, this equation tells us a new value for `x[i]` that is (hopefully) better than the previous guess.

For some problems, this updated value for `x[i]` might move the unknown too far and our approximate solution might start to diverge from the correct solution. In this case, the following line of code in the algorithm for calculating a new value for `x[i]` might provide greater stability:

```
x[i] = (1.0 - omega) * x[i] + \
       omega * (f[i] - sum)/A[i,i]
```

where `omega` is set to a value between 0.0 and 1.0. This change to the Gauss–Seidel iteration results in a new guess at `x[i]` that is equal to a weighted average of the old guess plus an updated guess. This approach is called under-relaxation, and it can help with stability at the cost of potentially slowing converge. Sometimes, the convergence is dramatically slower and 2–3 times as many iterations are required.

Similarly, one can imagine situations where we wish to try to move faster toward the solution. If the new guess for `x[i]` is really a much better guess, maybe we should try to move even further in that same direction. Using the same updated line of the algorithm as under-relaxation used, setting `omega` to a value greater than 1.0 can potentially accelerate convergence. When `omega` is set larger than 1.0, the method is called successive over-relaxation or SOR, for short. For the Gauss–Siedel example given above, setting `omega = 1.1` can reduce the number of iterations required for convergence by 1 to a total of seven iterations. Unfortunately, increasing `omega = 1.5` increases the number of iterations by a factor of 2 because we tend to overshoot and over-correct **x** each iterations. In summary, this simple move to the use of relaxation methods can help with either robustness of convergence or accelerate convergence, but the algorithm is now more complex and that complexity can be cause harm if care is not exercised. Before closing this section on iterative methods, it is helpful to look at the factors that impact whether or not these methods converge and the rate of convergence.

6.3.5 Convergence of Iterative Methods

The goal of the iterative methods described here is to determine an approximate solution to the problem $A \cdot x = f$. To examine the implications of the differences between the methods described previously, it is useful to add and subtract $I \cdot x$ from the right-hand side (note that I is the identity matrix, which is of the same size as A but just has ones on the diagonal and zeros everywhere else), giving: $I \cdot x + A \cdot x - I \cdot x = f$. This equation can be rearranged to give a potential iterative method: $x_{new} = f - (A - I) \cdot x_{old}$. It turns out that this is a really slow iterative method that should never be used.

We can use this same framework to write down the Jacobi iteration. We begin by decomposing A into a matrix D that just has the main diagonal terms from A with zero everywhere else, a matrix L that just has the values from A that are in the lower triangular section strictly below the main diagonal, and a matrix U that contains the values from A that are above the main diagonal. With these new matrices, we can rewrite $A = D + L + U$. An example of this decomposition for our distillation example is

$$
\begin{bmatrix} 0.9 & 0.3 & 0.1 \\ 0.1 & 0.5 & 0.2 \\ 0.0 & 0.2 & 0.7 \end{bmatrix} =
$$
$$
\begin{bmatrix} 0.9 & 0.0 & 0.0 \\ 0.0 & 0.5 & 0.0 \\ 0.0 & 0.0 & 0.7 \end{bmatrix} + \begin{bmatrix} 0.0 & 0.0 & 0.0 \\ 0.1 & 0.0 & 0.0 \\ 0.0 & 0.2 & 0.0 \end{bmatrix} + \begin{bmatrix} 0.0 & 0.3 & 0.1 \\ 0.0 & 0.0 & 0.2 \\ 0.0 & 0.0 & 0.0 \end{bmatrix}. \tag{6.15}
$$

Recalling that in the Jacobi iteration, all the off diagonal terms in A were effectively moved to the right side of the equation and we then divided by the diagonal terms of A, the Jacobi iteration can be written as

$$x_{new} = D^{-1}(f - (L + U) \cdot x_{old}).$$

Using the same strategy, the Gauss–Seidel iteration can be written as

$$x_{new} = (D + L)^{-1}(f - U \cdot x_{old}).$$

In either case, we are required to calculate the inverse of a matrix, D^{-1} or $(D + L)^{-1}$, which is normally the same computational cost as Gauss Eliminate (i.e., order n^3 operations) but is very inexpensive for the two matrices listed here because they are strictly diagonal or lower triangular (i.e., the same cost as backward substitution, order n^2 or less). As a result, each iteration is relatively inexpensive from a computational standpoint.

The rate at which these iterative methods converge depends on how well the *preconditioner*: D^{-1} or $(D + L)^{-1}$ for Jacobi and Gauss–Seidel, respectively, approximates A^{-1}. If the diagonal matrix, D, contains the largest terms in A, then D^{-1} is a good preconditioner and convergence is rapid. If the largest magnitude terms are not along the main diagonal, that is, not contained in **D**, then it is a poor preconditioner and convergence is unlikely.

The field of iterative methods for systems of linear equations is very broad, and the presentation of convergence rate here is very simplified. There are a number of iterative methods such as the conjugate gradient method, Krylov methods, and multigrid methods that are beyond the scope of this book. However, in all cases, the availability or absence of a good and inexpensive preconditioner has a significant impact on the performance of the method. Interestingly, one common precondition is to use Gaussian elimination but to throw away any small terms that arise during the computations. This is referred to as incomplete elimination and it provides a robust and inexpensive preconditioner for some problems. For more information on iterative methods, the interested reader is encouraged to read the following:

- Numerical Analysis by Burden and Faires [3]
- Iterative Methods for Solving Linear Systems by Greenbaum [4]
- A Multigrid Tutorial by Briggs *et al.* [5]

Problems

6.1 You have been hired by the EPA (Environmental Protection Agency) to estimate the concentration of PCBs (polychlorinated biphenyls) in the Great Lakes. In order to perform this analysis, you need to recognize that the quantity (in kg/year) of PCBs entering each lake must equal the quantity of PCBs leaving each lake (otherwise, the quantity in the lake would increase until it was infinite). Mathematically, we can write this as *in* = *out*. The quantity of PCBs in any river between the lakes can be calculated by multiplying the flowrate (in $\frac{km^3}{year}$) by the concentration (in $\frac{kg}{km^3}$).

Looking at the diagram below, we write a balance (*in* = *out*) on each lake. For Lake Superior,

$$180\frac{kg}{year} = Q_{SH} \cdot C_S = 72\frac{km^3}{year} \cdot C_S,$$

and for Huron, the balance is (note that the input is the sum of three streams):

$$630\frac{kg}{year} + 72\frac{km^3}{year} \cdot C_S + 38\frac{km^3}{year} \cdot C_M = 160\frac{km^3}{year} \cdot C_H.$$

Begin by deriving the five balances on the five lakes (note, two of them appear above). Then, solve the system of five linear equations (not by hand, but using one of the methods covered in this chapter) for the five unknowns (C_S, C_M, C_H, C_E, and C_O). Note that you will need to rearrange the equations so that the terms with unknowns all appear on one side of the equation. Report to the EPA the concentrations it is seeking.

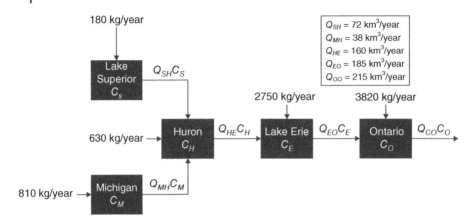

The EPA is also considering a bypass stream that would go directly from Lake Michigan to Lake Ontario with a flow rate of $20\frac{\text{km}^3}{\text{year}}$ in order to reduce the concentration of PCBs in Lake Michigan. This bypass does not change any of the existing flow rates, it would just be an additional flow out of Lake Michigan and into Lake Ontario. Report the potential impact of the bypass.

6.2 A new type of chair for a ski lift has been developed, and the manufacture has designed a simplified model (Figure 6.4) of the chair's behavior upon loading a group of individuals. Recall that the basic spring equation is

$$W = k \cdot x,$$

where W is the weight or force applied to the spring, k the spring constant, and x the displacement (or stretch) of the spring.

You have been hired to calculate the total displacement of the system of springs and weights shown in Figure 6.4. The properties of the system are as follows:

Parameter	Value
k_1	10,000.0 N·m
k_2	5,000.0 N·m
k_3	8,000.0 N·m
k_4	3,500.0 N·m
k_5	4,500.0 N·m
W_1	500.0 N
W_2	1,000.0 N
W_3	1,000.0 N

Figure 6.4 A system of weights and linear springs that model a new chair lift design for ski resorts.

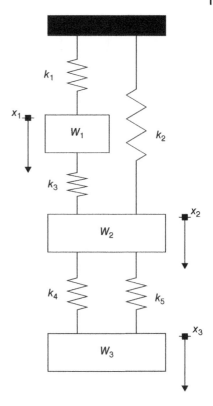

The first step is to derive a force balance on each weight. For the first weight, W_1, you need to consider every spring *touching the weight*, including the displacement and direction of force:

$$W_1 = k_1 \cdot x_1 - k_3 \cdot (x_2 - x_1)$$

and for the second weight:

$$W_2 = k_2 \cdot x_2 + k_3 \cdot (x_2 - x_1) - k_4 \cdot (x_3 - x_2) - k_5 \cdot (x_3 - x_2).$$

After deriving the force balance for the third weight, the system of equations can be written as a linear matrix problem:

$$\begin{bmatrix} k_1 + k_3 & -k_3 & 0.0 \\ -k_3 & k_2 + k_3 + k_4 + k_5 & -k_4 - k_5 \\ ? & ? & ? \end{bmatrix} \cdot \begin{bmatrix} x_1 \\ x_2 \\ x_3 \end{bmatrix} = \begin{bmatrix} 500.0 \\ 1000.0 \\ 1000.0 \end{bmatrix}. \quad (6.16)$$

If the linear matrix has been derived correctly, the terms along the main diagonal will all be positive, off-diagonal terms will be negative, and the matrix will be symmetric. Solve the linear matrix problem and determine the displacement of each weight (i.e., each skier) in *m*.

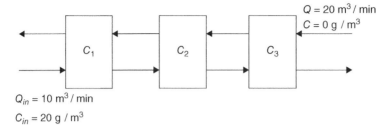

Figure 6.5 A three-stage, counter current cascade where all the lower streams have a flow rate of 10 m³/min and all upper streams have a flow rate of 20 m³/min. The feed stream with the compound of interest is fed into the first (left) stage at a concentration of 20 g/m³.

6.3 Counter current cascades are common in Chemical Engineering for the extraction of a substance from one stream into another stream. A three-stage counter current cascade is shown in Figure 6.5. Each stage in the cascade is basically a mixing tank, and we will assume that both streams leaving a stage have the same concentration (i.e., both the upper and lower stream *leaving* stage 1 have a concentration of C_1).

You have been hired by a local environmental engineering firm to analyze the effectiveness of the cascade shown below. You should begin by writing a mass balance on each stage because for each stage the total mass of the contaminant coming *in* must equal the mass of contaminant going *out*. In other words, grams per min of contaminant coming in equals grams per minute of contaminant going out. Looking at stage 1, there are two input streams, so the total contaminant coming in is

$$Total = Q_{in} \cdot C_{in} + 20 \ \text{m}^3/\text{min} \cdot C_2$$

If we include the total going out to this equation and replace the known variables, we get

$$\underbrace{(10 \, \text{m}^3/\text{min})(20 \, \text{g/m}^3) + (20 \, \text{m}^3/\text{min})(C_2)}_{\text{inputs in g/min}} =$$

$$\underbrace{(10 \, \text{m}^3/\text{min})(C_1) + (20 \, \text{m}^3/\text{min})(C_1)}_{\text{outputs in g/min}} \qquad (6.17)$$

and for the second stage balance, we get

$$(10 \ \text{m}^3/\text{min})(C_1) + (20 \ \text{m}^3/\text{min})(C_3) =$$
$$(10 \ \text{m}^3/\text{min})(C_2) + (20 \ \text{m}^3/\text{min})(C_2). \quad (6.18)$$

After deriving the mass balance for the third stage, the system of equations can be written as a linear matrix problem by moving all terms containing an unknown, C_i to the left side of the equal sign, and all terms without an unknown to the right side, giving:

$$\begin{bmatrix} -10-20 & 20 & 0.0 \\ 10 & -10-20 & 20 \\ ? & ? & ? \end{bmatrix} \cdot \begin{bmatrix} C_1 \\ C_2 \\ C_3 \end{bmatrix} = \begin{bmatrix} -200.0 \\ 0.0 \\ 0.0 \end{bmatrix}. \tag{6.19}$$

If the linear matrix has been derived correctly, the terms along the main diagonal will all be negative and off-diagonal terms will be positive. Solve the linear matrix problem and determine the concentration in each stage of the cascade in g/m^3.

6.4 Use the bobcatLU() and bobcatBS() functions to solve the linear matrix system:

$$\begin{bmatrix} 4.0 & -1.0 & 1.0 \\ 2.0 & 5.0 & 2.0 \\ 1.0 & 2.0 & 4.0 \end{bmatrix} \cdot \begin{bmatrix} x_1 \\ x_2 \\ x_3 \end{bmatrix} = \begin{bmatrix} 8.0 \\ 3.0 \\ 11.0 \end{bmatrix}. \tag{6.20}$$

Further, compare the solution to the solution obtained using numpy.linalg.solve(). In both cases, the solution should be $[1.0, -1.0, 3.0]$.

6.5 Use the bobcatLU Gaussian elimination algorithm on the following linear matrix system:

$$\begin{bmatrix} 1.0 & -1.0 & 2.0 & -1.0 \\ 2.0 & -2.0 & 3.0 & -3.0 \\ 1.0 & 1.0 & 1.0 & 0 \\ 1.0 & -1.0 & 4.0 & 3.0 \end{bmatrix} \cdot \begin{bmatrix} x_1 \\ x_2 \\ x_3 \\ x_4 \end{bmatrix} = \begin{bmatrix} -8.0 \\ -20.0 \\ -2.0 \\ 4.0 \end{bmatrix}. \tag{6.21}$$

The simple Gaussian elimination algorithm implemented in bobcatLU() fails with this matrix because during the elimination process, a zero is produced on the main diagonal, which in the best case triggers an error message stating that pivoting is required, and in the worst case causes a program crash.

Rewrite the bobcatLU() algorithm to use pivoting to avoid such difficulties. A description of a pivoting algorithm can be found in a number of numerical methods books including the book by Burden and Faires [3].

References

1 Zamir, M. (2016) *Hemo-Dynamics,* Biological and Medical Physics, Biomedical Engineering, Springer International Publishing, Heidelberg.
2 Fung, Y. (1984) *Biodynamics: Circulation,* Springer-Verlag, New York.
3 Burden, R. and Faires, J. (2001) *Numerical Analysis,* Brooks/Cole, Pine Grove, CA, 7th edn.
4 Greenbaum, A. (1997) *Iterative Methods for Solving Linear Systems, Frontiers in Applied Mathematics,* Society for Industrial and Applied Mathematics, Philadelphia, PA.
5 Briggs, W., Henson, V., and McCormick, S. (2000) *A Multigrid Tutorial,* Society for Industrial and Applied Mathematics, Philadelphia, PA, 2nd edn.

7

Regression

7.1 Motivation

Frequently in engineering, a series of measurements are taken, while one or more parameters for a system are varied. In some case, such as the sample data shown in Figure 7.1(a), the measured system parameter (also called the dependent variable) varies linearly with the parameter that is being systematically varied. For example, if the pressure in a rigid tank containing an ideal gas is measured while the temperature is changing, the relationship will be linear as long as the gas behaves like an ideal gas. In this example, pressure is the dependent variable and temperature is the independent (controlled) variable. The standard practice is to plot the data with the dependent variable (pressure) on the "y-axis" and the independent variable (temperature) on the "x-axis". The other possible case, which is much more common in real world, is that the measured variable changes nonlinearly as the controlled parameter is being varied. An example of this occurs if we measure the vapor pressure of water while varying the temperature of the liquid. Example data of such an experiment is shown in Figure 7.1(b). The careful observer will note that in the event nonlinear behavior is observed, one can simply reduce the range over which the independent variable is changed to recover a small region where the relationship is approximately linear. All curves are approximately linear if you zoom in far enough.

Data tends to be much more useful if we are able to obtain a mathematical expression that approximately matches or "fits" the data. If only two data points are available, then it is trivial to determine the equation for the line that passes through those two points. (It is extremely dangerous to do this because we typically do not know if additional data will be linear and fall close to this line.) If more than two data points are available, then all three points are extremely unlikely to fall on the same line so we need to develop an approach for obtaining the equation for a line that "best fits" the data.

This chapter covers linear regression in some detail, including the mathematics used to find the equation for a line that minimizes the square of the distance between the data and the linear regression line. Nonlinear regression, where data is fit with a nonlinear mathematical function that is specified by

Chemical and Biomedical Engineering Calculations Using Python®, First Edition. Jeffrey J. Heys.
© 2017 John Wiley & Sons, Inc. Published 2017 by John Wiley & Sons, Inc.
Companion Website: www.wiley.com/go/heys/engineeringcalculations_python

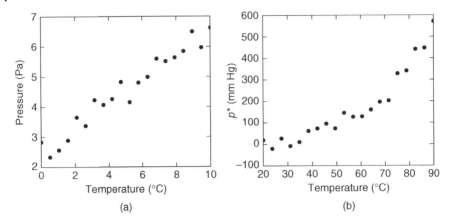

Figure 7.1 Example of two different data sets that might require regression to fit the data with either a line (a) or some other nonlinear function (b).

the user, is also covered using tools available in Scipy. However, before getting into the mathematics and programming, let us examine a common problem in vapor–liquid equilibria that requires regression.

7.2 Fitting Vapor Pressure Data

Vapor pressure is the pressure exerted by a pure liquid into the vapor phase. In other words, it is a measure of the volatility of the liquid, that is, a measure of how badly the liquid wants to be a vapor. If the vapor pressure reaches or exceeds the total pressure in the vapor phase, the liquid boils. If the vapor pressure of water in a pot on the stove exceeds approximately 760 mm Hg, that is, atmospheric pressure, then the water boils. The vapor pressure of water at 100 °C is, of course, 760 mm Hg. Figure 7.1(b) shows a typical set of vapor pressure measurements for water over a range of temperatures.

According to the Clausius–Clayperon equation, the relationship between vapor pressure, p^*, and temperature, T, is an equation of the form:

$$\ln(p^*) = c_1 \left(\frac{1}{T} \right) + c_2,$$

where c_1 and c_2 are constants that must be determined from experimental measurements. Looking at the data show in Figure 7.1(b) it may not be apparent that we could ever use linear regression to fit the data, but the Clausius–Clapeyron relationship shows us the way. Recalling that the equation for a line is

$$y = c_1 \cdot x + c_2,$$

we notice that if we plot $\ln(p^*)$ (instead of just p^*) and if we plot $\frac{1}{T}$ instead of T, the data should be approximately linear, allowing us to fit the data with a

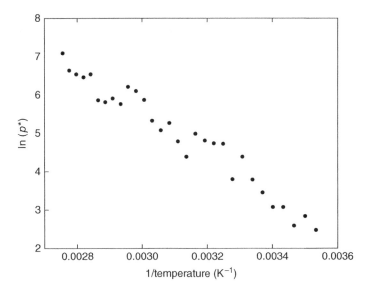

Figure 7.2 When the natural log of vapor pressure, $\ln(p^*)$, is plotted against the inverse temperature, $1/T$, then the points fall on a straight line and linear regression can be used to fit the data.

line and determine c_1 and c_2. When vapor pressure data, such as that shown in Figure 7.1(b), is replotted after taking the natural log of the vapor pressure and the inverse of temperature, the data falls on an approximately straight line as shown in Figure 7.2. Once the constants are known, we can use the Clausius–Clapeyron relationship to determine the vapor pressure at any temperature we desire or, equivalently, to determine the temperature at which a desired vapor pressure can be achieved.

7.3 Linear Regression

To begin the regression process, we need data. For now, assume that we have n data points that consist of one independent variable, x, and one dependent variable, y, so there are n data points: (x_i, y_i), $i = 1, \dots, n$. The goal is to obtain an equation for a polynomial, $y = p(x)$, that approximately matches the data. The first step is the selection of a measure that tells us how well the polynomial matches the data. For example, we may want to minimize the absolution value of the distance between the polynomial, $p(x)$, and the data, y_i. If this is the goal, then we need to minimize:

$$E_{abs} = \sum_{i=1}^{n} |y_i - p(x_i)|. \tag{7.1}$$

In practice, it is difficult to minimize E because the derivative of the absolute value function is not defined at the origin. Least-squares regression is an easier and much more common choice. It is based on minimizing:

$$E_{ls} = \sum_{i=1}^{n} (y_i - p(x_i))^2. \tag{7.2}$$

In order to understand the process of minimizing E_{ls}, let us assume for simplicity that we are interested in linear polynomials, that is, $p(x_i) = c_1 x_i + c_0$. (It is relatively straightforward to extend this analysis to higher-order polynomials, but this is often unnecessary because helpful software has been developed to automate the process.)

The goal of linear regression analysis is to determine the two unknowns in the equation for the line: c_0 and c_1 that minimize E_{ls}. As we probably learned in one of our calculus courses, the local minimum of a function occurs when the derivative is equal to zero. Further, since E_{ls} is a quadratic function, the local minimum is also the global minimum. Taking the derivative of E_{ls} with respect to the two unknowns gives

$$\frac{\partial E_{ls}}{\partial c_0} = -2 \sum_{i=1}^{n} (y_i - c_1 x_i - c_0) = 0,$$

$$\frac{\partial E_{ls}}{\partial c_1} = -2 \sum_{i=1}^{n} ((y_i - c_1 x_i - c_0) \cdot x_i) = 0.$$

Simplifying these equations (note that a constant can be factored out of a sum) gives

$$\sum_{i=1}^{n} (y_i) = c_1 \sum_{i=1}^{n} (x_i) + n \cdot c_0, \tag{7.3}$$

$$\sum_{i=1}^{n} (y_i x_i) = c_1 \sum_{i=1}^{n} (x_i^2) + c_0 \sum_{i=1}^{n} (x_i), \tag{7.4}$$

which are commonly referred to as the normal equations. Recall that x_i and y_i are *known* or given by the data. The only unknowns are c_0 and c_1. In terms of these unknowns, the normal equations are a linear system of equations and can be rewritten as a matrix problem:

$$\begin{bmatrix} n & \sum_{i=1}^{n}(x_i) \\ \sum_{i=1}^{n}(x_i) & \sum_{i=1}^{n}(x_i^2) \end{bmatrix} \cdot \begin{bmatrix} c_0 \\ c_1 \end{bmatrix} = \begin{bmatrix} \sum_{i=1}^{n}(y_i) \\ \sum_{i=1}^{n}(x_i y_i) \end{bmatrix}. \tag{7.5}$$

Fortunately, we learned how to solve linear matrix problems in the previous chapter.

A Python script that performs linear regression on eight data points, which are given near the start of the script, is shown below.

```
import numpy as np
import numpy.linalg as nl
import pylab

n = 8
x = np.array([0.1, 1.43, 2.86, 4.29, 5.71,
              7.14, 8.57, 9.95])
y = np.array([2.33, 2.81, 3.84, 4.41, 4.31,
              5.65, 5.68, 6.80])
if ((n!=x.size) or (n != y.size)):
    print("Error:inconsistent number of data points")

Xsum = np.sum(x)
Ysum = np.sum(y)
XYsum = np.sum(np.dot(x,y))
X2sum = np.sum(np.dot(x,x))
A = np.array([[n, Xsum], [Xsum, X2sum]])
f = np.array([Ysum, XYsum])
c = nl.solve(A, f)

yLR = c[0] + c[1]*x
pylab.plot(x,y,'o',x,yLR)
pylab.xlabel('x')
pylab.ylabel('y')
pylab.show()
```

The script begins by constructing two numpy arrays that contain the data. One array, **x**, contains the independent variable and the other array, **y**, contains the dependent variable data. The script then checks to make sure that the arrays are the same size. A useful numpy function, numpy.**sum**(), is used to sum the entries in a vector. These sums are stored in different variables for use later in constructing the matrix, **A**, and the right-hand side, **f** as defined in equation 7.5. The linear system is solved using the solver included with numpy. In order to plot the regression line, the solution to the linear problem, which contains the coefficients for the linear polynomial fit to the data, is used to calculate the approximate value of the dependent variable at every independent variable point. The original data for the example problem is shown in Figure 7.3(a), and the data with the linear regression curve is shown in Figure 7.3(b). Plotting data as separate points and polynomial fits to the data as a solid line is standard practice and highly recommended.

The data for the previous example was obtained by first selecting evenly distributed points from a straight line: $y = 0.4x + 2.5$ and then adding random noise to the data so that it did not all fall on a straight line. Interestingly,

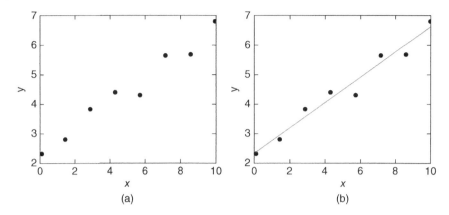

Figure 7.3 (a) Sample data used in regression analysis and (b) the same data with the calculated regression line.

the regression line obtained using least-squares linear regression was $y = 0.4285x + 2.33$. Because only eight points were used and the random noise was not evenly distributed over such a small number of points, the original line was not completely recovered. Additional testing showed that the slope could be reliably recovered (approximately two significant digits) with about 20 data points, but recovering an accurate value for the intercept was more difficult and required at least 100 data points to obtain approximately two significant digits.

Quadratic Regression

Repeating the derivation of the normal equations and linear matrix problem associated with the least-squares regression of a quadratic polynomial: $y_i = c_2 x_i^2 + c_1 x_i + c_0$ gives

$$\begin{bmatrix} n & \sum_{i=1}^{n}(x_i) & \sum_{i=1}^{n}(x_i^2) \\ \sum_{i=1}^{n}(x_i) & \sum_{i=1}^{n}(x_i^2) & \sum_{i=1}^{n}(x_i^3) \\ \sum_{i=1}^{n}(x_i^2) & \sum_{i=1}^{n}(x_i^3) & \sum_{i=1}^{n}(x_i^4) \end{bmatrix} \cdot \begin{bmatrix} c_0 \\ c_1 \\ c_2 \end{bmatrix} = \begin{bmatrix} \sum_{i=1}^{n}(y_i) \\ \sum_{i=1}^{n}(x_i y_i) \\ \sum_{i=1}^{n}(x_i^2 y_i) \end{bmatrix} \tag{7.6}$$

7.3.1 Alternative Derivation of the Normal Equations

Recall that the goal in polynomial regression analysis is to determine the polynomial coefficients that give the optimal fit to n data points. This requires that the number of data points be greater than the number of coefficients in the polynomial. The polynomial will have the form: $y_i = c_0 + c_1 x_i + c_2 x_i^2 + \cdots$. Since $i = 1, \ldots, n$, we can think of this polynomial at each data point being written as a linear matrix problem: $\mathbf{Ac} = \mathbf{y}$ where

$$
\mathbf{A} = \begin{bmatrix} 1 & x_1 & x_1^2 & \cdots \\ 1 & x_2 & x_2^2 & \cdots \\ \vdots & & & \ddots \\ 1 & x_n & x_n^2 & \cdots \end{bmatrix}
\tag{7.7}
$$

Notice that $\mathbf{A}$ and $\mathbf{y}$ are both given by the data that is being fit. In regression, the goal is to minimize

$$
E_{ls} = \|\mathbf{y} - \mathbf{Ac}\|^2,
\tag{7.8}
$$

which is the L^2-norm of the residual squared. E_{ls} is minimized when the derivative is zero, which leads to

$$
\frac{\partial E_{ls}}{\partial c} = 2\|\mathbf{y} - \mathbf{Ac}\|\mathbf{A} = 0.
$$

Upon rearrangement [1], we obtain the normal equations in matrix form:

$$
(\mathbf{A}^\mathsf{T}\mathbf{A})\mathbf{c} = \mathbf{A}^\mathsf{T}\mathbf{y}.
\tag{7.9}
$$

The result is the same linear matrix problem that was derived previously using the more common approach.

7.4 Nonlinear Regression

The fitting of data by a nonlinear function through a least-squares minimization process can be difficult because the process leads to a system of nonlinear equations that must be solved. The solving of nonlinear equations is the focus of the next chapter, but a few words on the process are important here. First, nonlinear regression requires that the user supply a "guess" for the values of the unknown parameters in the function that is being fit to the data. If data is being fit to a function like $a \sin(\pi x/p)$, where x is the independent variable, then the user must supply an initial guess for the amplitude, a, and the period, p, of the data. The nonlinear solution process tends to be much more likely to converge to a function that optimally fits the data in a least-squares norm (equation 7.1) if the initial guesses for the function parameters are relatively close. The second factor that impacts the solution process is the accuracy of the data. Data with significant levels of noise is unlikely to lead to a nonlinear regression solution for highly nonlinear functions.

A relatively robust nonlinear least-squares regression routine is included with the Scipy library, it is the `scipy.optimize.curve_fit()` function.

This function is based on the Levenberg–Marquardt algorithm [2] for nonlinear regression. The call to this function has the form:

```
scipy.optimize.curve_fit(func, xdata, ydata, p0, sigma)
```

and the following parameters are passed into this function when it is called:

func: This is a user-defined function that is declared before curve fit is called. The function must take the independent data as input (i.e., xdata) and the values for the different unknown parameters that are being determined in the fitting process. The function must return a vector, **y**, containing the dependent variable. The curve_fit function changes the values of the unknown parameters to minimize the difference between the function return values and the dependent variable data, ydata.

xdata: a vector of the independent variable data points.

ydata: a vector of the dependent variable data points. The **y**-values returned by the function (func) are compared to this data using the least-square norm (equation 7.1).

p0: a Python list containing guesses for the unknown function parameters. A default value of 1 is used for all parameters if no initial guess is provided. Use of the default value is discouraged.

sigma: an *optional* vector that is used to provide relative weights for the least-squares processes. If the goal is to fit some of the data points more closely than other data, a larger weight can be applied to those data points. This vector is rarely provided.

The use of the `curve_fit()` function is illustrated through the Python script below.

```python
import numpy
from scipy.optimize import curve_fit
import pylab


n = 20
# Antoine coefficients for water from Wikipedia
A = 8.07131
B = 1730.63
C = 233.426


# Build some fake data: temperature, x, versus
# vapor pressure,  y, data
x = numpy.linspace(20,90,num=n)
error = numpy.random.rand(n)
y = numpy.zeros(n)
for i in range(n):
```

```
      y[i] = A - B/(x[i]+C)
      y[i] = (10**y[i])+50*(error[i]-0.5)

# Function for Antoine's equation - used in call
# to curve fit below.
def antoine(temp, a, b, c):
    n = temp.size
    p = numpy.zeros(n)
    for i in range(n):
        p[i] = 10**(a - b/(temp[i]+c))
    return p

# Guesses for the Antoine coefficients and curve_fit call
params = [10, 2000, 200]
popt, pcov = curve_fit(antoine, x, y, p0=params)

# calculate the dependent variable for  plotting the curve
yfit = antoine(x, popt[0], popt[1], popt[2])
# plot data as points and fit as a line
pylab.plot(x,y,'o',x, yfit)
pylab.xlabel('temperature ($^oC$)')
pylab.ylabel('$p^*$ (mm Hg)')
pylab.show()
```

This function begins by generating some "fake" data that is curve fit later in the script. Random noise is added to the fake vapor pressure data. The nonlinear function with unknown parameters is defined in the function `antoine(temp,a,b,c)`. This function is passed a vector of temperatures and estimates for the three unknown parameters: *a*, *b*, and *c*. The function returns the vapor pressure, p_i, at a given temperature, T_i, using a function of the form:

$$p_i = 10^{a+b/(T_i+c)}, \tag{7.10}$$

which is known as Antoine's equation. The `curve_fit()` function repeatedly calls `antoine()` with different estimates for the parameters (*a*, *b*, and *c*) in an effort to better fit the dependent variable data in the vector y.

The `scipy.optimize.curve_fit()` function returns a tuple with two numpy arrays. The first item in the tuple is a vector (stored in `popt` in the example above) containing the optimal parameter values for fitting the function to the data. The second item in the tuple is an array (stored in `pcov` in the example above) containing the covariance matrix associated with the optimal parameter values. At the end of the script, the original fake data and the `curve_fit()` result, if one is obtained, are plotted. Plotting the function (or curve) that optimally fits the data requires a set of (x, y) values that fall on the curve. In the example above, the *x*-data is also used for plotting the curve,

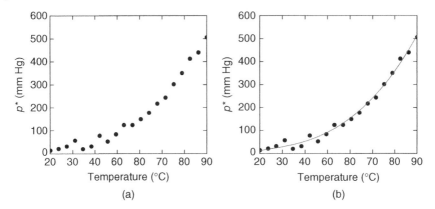

Figure 7.4 (a) Sample data used in nonlinear regression analysis and (b) the same data with the nonlinear regression line from `curve_fit()`. $p^* = 10^{8.1-1719/(232+T)}$.

but new y-data is obtained using the optimal parameter values from the popt vector and the function that was used in the fitting (`antoine()`).

A sample result using the script above is shown in Figure 7.4. Repeated experimentation with the script and different sets of fake data provided some interesting observations. First, the Antoine parameters from `curve_fit()` often differed from the actual parameters by 10% or more whenever significant quantities of noise were added to the fake data. Second, the `curve_fit()` function often failed to converge to a least-squares minimizer if the error level in the data exceeded roughly 10%. Adding small amounts (0.1%) of noise to the data gave accurate values for the Antoine parameters and always led to convergence by `curve_fit()`, but larger amounts of noise gave poor results or no results. Finally, large quantities of data (50 or more points) gave better results than small quantities of data (10 or fewer points).

7.4.1 Lunar Disintegration

The book Seveneves by Neal Stephenson is a science fiction novel that begins with an event, caused by an unknown "agent", that causes the moon to break apart into seven large fragments. Relatively quickly, scientists in the book recognize that the seven large fragements will collide with each other and cause the large fragments to break apart into smaller fragments. As more fragments are formed, the potential for collision and further fragmentation increase. This system exhibits a classic exponential growth curve – growth (an increase in the number of fragments) in one generation causes faster growth (a faster rate of new fragment formation) for the next generation. In the book, scientist recognize that this fragmentation process would continue until the original moon was broken apart into an incredibly large number of small fragments. Some of

Table 7.1 Estimates of the number of fragments of the former moon as a function of the number of days since the initial events.

Day:	0	7	28	100	200	300
Fragments:	7	8	20	350	12,000	500,000

the fragments would reach earth due to the earth's gravity, and these fragments would impact the surface or burn up in the atmosphere. Due to the incredibly large number of fragments that the former moon would produce, the earth's surface and atmosphere would heat up beyond the point where life of any form would survive.

The text of the book does not clearly state the exact number of fragments that the moon has been reduced to as a function of time, but through a careful analysis of the text, the data in Table 7.1 has been estimated and includes the number of fragments as a function of the number of days from the initial event (day 0). Our goal is to fit the fragment number versus day data with an exponential curve of the form:

$$f = a \cdot e^{x/\tau},$$

where f is the number of fragments, x is the day, and a and τ are unknown parameters.

The Python script below uses the `curve_fit()` function to determine the values for a and τ that result in the exponential curve to best fit the fragmentation data.

```
import numpy
from scipy.optimize import curve_fit
import matplotlib.pyplot as plt

day = numpy.array([0, 7, 28, 100, 200, 300])
chunks = numpy.array([7, 8, 20, 350, 12000, 500000])

# Function for Breakup equation
# used in call to curve_fit below.
def breakup(x, a, tau):
    f = a*numpy.exp(x/tau)
    return f

# Guesses for the Breakup coefficients
params = [7.0, 30.0]

# call curve_fit, returns fitting params
```

```
# and covariance of params, ignored with_
fit,_ = curve_fit(breakup, day, chunks, p0=params)

print(fit)

moreDays = numpy.arange(0,700)
yfit = breakup(moreDays, fit[0], fit[1])
# plot data as points and fit as a line
plt.subplot(121)
plt.semilogy(day,chunks,'o', moreDays, yfit)
plt.xlabel('Day after the event')
plt.ylabel('Number of chunks')
plt.subplot(122)
plt.plot(day,chunks,'o', moreDays, yfit)
plt.xlabel('Day after the event')
plt.savefig('seveneves700.png')
```

The script begins by storing the fragmentation and day data in numpy arrays and then the exponential function that is to fit the data is defined. The `breakup()` function is called by `scipy.optimize.curve_fit()` a few lines below, and the `breakup()` function needs to be written to receive *three* arguments: a numpy *vector*, x, and two scalar parameters to be fit, a and `tau`. A possibly more robust approach to writing the `breakup()` function is to write a **for** loop to iterate through the values in x, but the more compact form shown here is possible due to the use of the `numpy.exp()` function.

As described earlier, the `curve_fit()` function returns two data structures: (1) the optimal values of the fitting parameters, stored in the variable `fit` above, and (2) covariance values for the parameters. For this example, we are not interested in the covariance values for the parameters, so we temporarily store them in the variable "_", which is a standard variable in Python for temporarily storing something that you do not plan to use. The result of fitting the fragmentation data with an exponential curve is shown in Figure 7.5. The exponential function fitting the data is

$$f = 6.92 \cdot e^{x/26.8}.$$

Note that a `semilogy()` plot is used due to the large variation in the number of fragments over time.

The final section of the code generates a plot that consists of two subplots, and in both cases, the exponential fragmentation curve is extrapolated beyond the data that is available. The data ends at day 300, but the curves shown in Figure 7.6 extrapolate the curve out to day 700. Extrapolation requires that we construct a new vector containing days out to 700, and this is achieved using the `numpy.arange()` function. The `subplot()` function is then used to

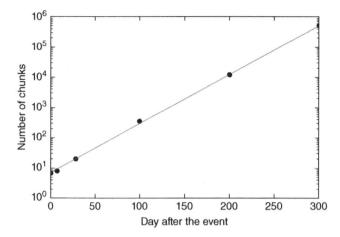

Figure 7.5 Fragmentation versus day after the event data is shown as descrete points, and the optimal exponential curve fitting the data is shown as a solid line.

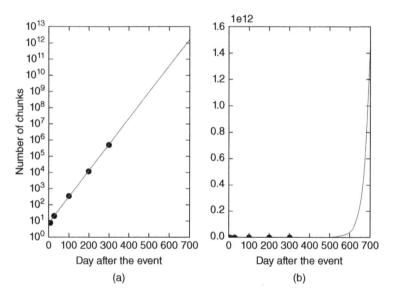

Figure 7.6 Fragmentation versus day after the event data is shown as descrete points, and the optimal exponential curve fitting the data is extrapolated out to 700 days and shown as a solid line.

generate a plot with two subplots, the subplot on the left uses a log-scale for the y-axis (using `semilogy`) and the subplot on the right uses a regular, linear y-axis to emphasize the exponential bend in the number of fragments as day 700 is approached.

7.5 Multivariable Regression

The linear and nonlinear regression problems that have been examined thus far in the chapter all have a single independent variable that is changing. For example, the vapor pressure was only a function of temperature. The number of lunar fragments was only a function of time (or days) since the initial fragmentation event. In most engineering problems, our goal is a fundamental understanding of a system and that means trying to change a single variable and observing the response of the dependent variable. However, sometimes this is impossible and there are multiple variables independently changing. For example, my heart rate is a function of activity level and caffeine consumption and it may not be possible to vary activity and caffeine consumption separately. The focus of this final section on regression is fitting, for example, an output variable, z, that depends on multiple input variables, x and y, with a function of the form: $z = a \cdot x + b \cdot y + c$. This example function has two independent variables and it is *linear*. It is also possible to fit with a function that has multiple independent variables and is nonlinear (e.g., $z = a \cdot x + b \cdot y + c \cdot x \cdot y + d \cdot x^2 + e \cdot y^2 + f \cdot x^2 \cdot y^2$). However, caution should be exercised to avoid fitting with too many parameters. If we were to fit with a function that had >10 parameters, we should never attempt this unless a large (>100) quantity of data is available, and even then, nonlinear terms may not be necessary.

To illustrate multivariable regression, imagine that data is available (the data can be found in the Python script below), and we need to fit the yield, y, of a chemical reaction with a function of the form: $y = a \cdot T + b \cdot C + d$, where T is the temperature and C the concentration. The Python script below begins by constructing `numpy.array()` vectors to hold the data. The regression or `curve_fit()` function below requires that all the data associated with the independent variables (T and C) be stored in a single array with each column holding the values for a different variable. The vectors holding the temperature and concentration data are combined into a single array using the `numpy.stack()` function. Next, the function to be fit to the data using least-squares regression is defined. The function contains three unknown parameters to be fit, a, b, and d, and initial guesses for the values of the parameters should be provided to the `curve_fit()` function, but because the function is linear, the guesses do not need to be of high quality. The Python code for the full process is below.

```python
import numpy
from scipy.optimize import curve_fit
import matplotlib.pyplot as plt

temp = numpy.array([150, 160, 170, 180,
                    150, 160, 170, 180,
                    150, 160, 170, 180,
```

```
                          150, 160, 170, 180])
conc = numpy.array([[40., 40., 40., 40.,
                      50., 50., 50., 50.,
                      60., 60., 60., 60.,
                      70., 70., 70., 70.])
x = numpy.stack((temp,conc),axis=1)
fracYield = numpy.array([70., 72., 74., 77.,
                          64., 66., 69., 71.,
                          49., 55., 57., 58.,
                          46., 48., 53., 55.])

def rxnFit(x, a, b, d):
    f = a*x[:,0] + b*x[:,1] + d
    return f

# Guesses for the Breakup coefficients
params = [1.0, 1.0, 1.0]

# call curve_fit, returns fitting params
# and covariance of params, ignored with_
fit,_ = curve_fit(rxnFit, x, fracYield, p0=params)

print(fit)
fitYield = rxnFit(x,fit[0],fit[1], fit[2])

bar_width = 0.2
index = numpy.arange(len(fracYield))
plt.bar(index,fracYield,bar_width)
plt.bar(index+bar_width,fitYield,bar_width,color='r')
```

It is very difficult to visualize or plot the fit of the function to the original data, especially in cases with a large number of independent variable. For the example above, a simple bar chart is constructed with pairs of bars as shown in Figure 7.7. The darker bar on the left is the original yield data for each sample, and the lighter bar on the right is the yield from the fitting function at the same temperature and concentration. If all pairs of bars have a similar height, that is, similar values, the fitting function agrees well with the data. For larger data sets, even the generation of a simple bar chart like this one is impossible and more sophisticated approaches are required for assessing the quality of the fit.

7.5.1 Machine Learning

The field of machine learning is exploding in popularity because computers have enabled the automated gathering and assembly of large data sets. For example, in chemical processing, it is possible to assemble a large data set

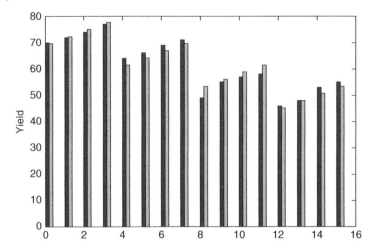

Figure 7.7 A bar chart showing pairs of bars for each experimental condition (temperature and pressure). The darker bar on the left is the yield from the original data, and the lighter bar on the right is the least-squares regression fit at the same conditions.

that holds the conditions (temperature, pressure, valve position, concentration, opacity, humidity, etc.) at a large number of locations throughout a chemical processing facility. Another example is complete records of each consumer's purchases at a grocery store over a multiyear period. The field of machine learning is focused on the development of algorithms that can fit these large data sets in order to predict outputs. For the chemical processing example, the output might be the purity of the final product, and the machine learning algorithm can predict purity based on a full set of input conditions, including the identification of the most important input conditions.

Probably, the most fundamental tool used in machine learning is multivariable regression. In the field of machine learning, the term "feature" is used instead of "variable", and the term "weight" is used instead of "parameter", but many of the basic ideas are the same as the multivariable regression problem solved above. To illustrate the connection between machine learning and multivariable regression, let us consider one of the most famous problems in machine learning: numerical character recognition. This problem involves scanning a single digit (0–9) from a hand-written number, thus converting it into an image or set of pixels. Then, based on the color (white, gray, or black) of the pixels, determine what number (0–9) is shown in the image. This is a pretty simple task for a human but very challenging for a computer. The approach used in machine learning is to start with a large collection (>1000) of images showing digits (0–9) that have already been identified by a person. The color of each pixel in the image is represented using a vector of numbers, **x**, and machine learning algorithms use regression to find a vector of weights, **w** such that $\mathbf{x} \cdot \mathbf{w} = y$ where y is the previously identified number. This is just a linear regression problem. We normally think of linear regression as: given a set of points, (x, y) find the parameters (m, b) of a line so that the line $y = m \cdot x + b$ is a good fit for determining y for any given x. The character recognition problem is very similar: find **w** so that $\mathbf{x} \cdot \mathbf{w} = y$ for any given vector of pixel colors, **x**.

Current challenges in machine learning include the challenge of getting a computer to answer a question. This problem usually starts with the process of converting every word in a language into a number. Then, we wish to find the set of weights, **w** such that $\mathbf{x} \cdot \mathbf{w} = y$ where **x** is the vector of numbers that represent the words in the question and y is the answer! If we have a large set of questions and answers (i.e., a large set of questions: **x** and answers y), it may be possible to find the desired weights (**w**). Finally, it is important to note that the field of machine learning significantly extends beyond multivariable linear regression. For the interested individual, the Machine Learning course through Coursera (https://www.coursera.org/learn/machine-learning) is one opportunity to learn more.

Problems

7.1 Look up the average monthly high temperature for your hometown. Starting with April, list the average monthly high temperature for the next 12 months. Then, fit this data with a function of the form:

$$a \cdot \sin(\pi x / p) + b$$

where x is the month (starting with April as zero and continuing to March, month 11) and a, b, and p are unknown parameters that are to be determined to fit the average monthly high in a least-squares minimization.

7.2 You have been hired by NASA to perform a regression analysis on the GLOBAL Temperature Index data set (original data: http://data.giss.nasa .gov/gistemp/tabledata_v3/GLB.Ts+dSST.txt). Using the data in the table below, complete a linear ($y = a + b \cdot x$), quadratic ($y = a + b \cdot x + c \cdot x^2$), and exponential ($y = ae^{bx}$) regression analysis of the average global temperature (y) versus year (x). Use of the `curve_fit()` function from the Scipy library is acceptable to NASA. Comment on the quality of the fit for the various regression options considered.

Year	Average global temperature (°C)	Approximate number of pirates
1890	13.8	20,000
1900	13.7	18,000
1910	13.7	16,000
1920	13.6	15,000
1930	13.8	9000
1940	13.9	5000
1950	14.0	3000
1960	14.0	2000
1970	14.0	1100
1980	14.0	400
1990	14.2	20
2000	14.4	17
2010	14.6	15

It has been suggested that the rise in global average temperature is due to the decline in the number of pirates. You have also been asked to perform a linear and quadratic regression analysis on the approximate number of pirates in the world (y) versus year (x). Finally, comment on the correlation between the approximate number of pirates in the world versus the average global temperature data.

7.3 You have been hired by the publisher of a Chemical Engineering reference manual to perform a regression analysis on Heat Capacity of solid Hydrogen Bromide data [3]. Using the data in the table below, complete a linear ($y = a + b \cdot x$), quadratic ($y = a + b \cdot x + c \cdot x^2$), and cubic ($y = a + b \cdot x + c \cdot x^2 + d \cdot x^3$) regression analysis of the Heat Capacity (y) versus temperature (x). Use of the `curve_fit()` function from the Scipy library is acceptable to the publisher. Comment on the quality of the fit for the various regression curves.

Temperature (K)	Heat capacity (cal/(mol · K))
118.99	10.79
120.76	10.8
122.71	10.86
125.48	10.93
127.31	10.99

Temperature (K)	Heat capacity (cal/(mol · K))
130.06	10.96
132.41	10.98
135.89	11.03
139.02	11.08
140.25	11.1
145.61	11.19
153.45	11.25
158.03	11.4
162.72	11.61
167.67	11.69
172.86	11.91
177.52	12.07
182.09	12.32

The final report to the publisher should include figures showing each of the regression curve fits, the equation for the fitting curve, and your recommendation for the fitting equation that best balances quality of fit with simplicity (i.e., fewer fitting parameters).

7.4 There is much debate about the proper approach for characterizing the rheology of blood. A particularly important aspect of this problem is establishing a relationship between the *shear stress* and *strain rate* for blood. The shear stress is a measure of the shear force per area applied to blood. Imagine a person sliding on a waterslide, the shear stress is the force per area that the person is applying to the water underneath them. The strain rate represents the rate (or time period) at which the fluid deforms. The magnitude of the deformation is usually normalized by the original size (or thickness) of the fluid so that the strain rate has units of "per time".

You have been hired by the Red Cross to fit strain rate and shear stress data from a blood sample to two different models, and then assess the quality of the models for fitting the data. The rheological data is from [4].

Strain rate (1/s)	Shear stress (dynes/cm^2)
1.5	12.5
2.0	16.0
3.2	25.2

Strain rate (1/s)	Shear stress (dynes/cm^2)
6.5	40.0
11.5	62.0
16.0	80.5
25.0	120
50.0	240
100	475

The first model proposed by the Red Cross is the linear (or Newtonian) model:

$$\tau = \mu \dot{\gamma},$$

where τ is the shear stress and $\dot{\gamma}$ the strain rate. The second model is the power law model:

$$\tau = k \cdot \dot{\gamma}^n.$$

Use least-squares regression (e.g., the `curve_fit()` function in Scipy) to determine the optimal value(s) for the unknown parameter(s) in each model (i.e., μ in the linear model or k and n in the power law model). Generate a plot showing the data and the optimal regression curve for each model. Finally, assess the quality of the fit in your report to the Red Cross.

7.5 Possibly the most fundamental relationship in ecology is that the number of unique species, S, increases as the area, A, of a region increases. This relationship has been quantitatively observed in numerous studies over the past 200 years. The relationship is normally of the form:

$$S = c \cdot A^z,$$

where c and z are unknown constants that depend on the ecological system being studies and they must be determined from experimental data. The value for z is a positive real number less than 1.0.

You have been hired as a consultant by the State of California to study the species–area relationship for endemic vascular plant species for partially isolated subregions of the state. Data on the number of species in each area was collected and published previously [5].

Location	Area (mi^2)	Species
Tiburon Peninsula	5.9	370
San Francisco	45	640

Location	Area (mi²)	Species
Santa Barbara area	110	680
Santa Monica Mountains	320	640
Marin County	529	1060
Santa Cruz Mountains	1386	1200
Monterey County	3324	1400
San Diego County	4260	1450
California Coast	24,520	2525

Fit the species versus area data to the power law relationship given above using regression to determine c and z for this particular system. An examination of similar studies shows that c is often of the order 10–1000 and z is frequently between 0.2 and 0.4.

On the basis of the regression analysis, the State of California has asked you to analyze the impact of the growth of its three major cities: Los Angeles, San Diego, and San Francisco. Specifically, if California loses 1% of the total land in the state to the development of the cities every year, approximately how many species will become extinct next year?

Hint: your supervisor recommends plotting the species versus area data and curve fit using a "semilogx()" plot in matplotlib.

7.6 The World Health Organization (WHO) publishes data on average height versus age for both females and males. Select data is summarized in the table below (age is given in month and height is given in cm).

Month:	0	12	36	60	96	144	168	192	216
Female:	49.1	74.0	95.1	109.4	126.6	151.2	159.8	162.5	163.1
Male:	49.8	75.8	96.1	110.0	127.3	149.1	163.2	172.9	176.1

You have been contracted by WHO to fit the data for females to the following growth function:

$$h = \frac{a}{1 + b \cdot e^{-c \cdot t}},$$

where h is the height of females in cm, t the age in months, and a, b, and c the unknown parameters to be determined using least-squares regression on this nonlinear function. Your report to WHO should include a plot of the height versus age data along with the optimal growth function to fit the data.

Finally, WHO is considering the potential for a single linear function to fit the data for both boys and girls, using a function of the form:

$$h = a \cdot t + b \cdot g + c,$$

where h is the height of the individual, g the gender (you need to propose an convervsion from gender, male or female, into a number), and a, b, and c unknown parameters to be determined using least-squares regression. You should comment on the quality of the fit for this function relative to the previous growth function for the data provided.

References

1 Anton, H. and Rorres, C. (2005) *Elementary Linear Algebra*, John Wiley & Sons, Inc., New York, 9th edn.

2 Moré, J. (1978) The Levenberg-Marquardt algorithm: implementation and theory, in *Numerical Analysis, Lecture Notes in Mathematics*, vol. 630 (ed. G. Watson), Springer-Verlag, Berlin Heidelberg, pp. 105–116.

3 Giaugue, W.F. and Wiebe, R. (1928) The heat capacity of hydrogen bromide from $15k$ to its boiling point and its heat of vaporization. The entropy from spctroscopic data. *J. Am. Chem. Soc.*, **80**, 2193–2202.

4 Palladino, J.L. and Davis, R.B. III (2012) Biomechanics, in *Introduction to Biomedical Engineering* (eds J.D. Enderle and J.D. Bronzino), Academic Press, Burlington, MA, pp. 134–218, 3rd edn.

5 Johnson, M.P., Mason, L.G., and Raven, P.H. (1968) Ecological parameters and plant species diversity. *Am. Nat.*, **102**, 297–306.

8

Nonlinear Equations

8.1 Introduction

Nonlinear equations are frequently encountered in diverse areas of engineering, including chemical reaction rates, phase equilibria, fluid distribution systems, and material deformation at large strain. One important nonlinear equation that will be used as an example problem in this chapter is the Soave–Redlich–Kwong (SRK) nonideal equation of state. This equation relates the pressure and temperature of a gas to its specific volume (i.e., the volume per mole of material). The SRK equation of state can be written as

$$P = \frac{RT}{\hat{V} - b} - \frac{\alpha a}{\hat{V}(\hat{V} + b)}, \tag{8.1}$$

where P is the absolute pressure, T the absolute temperature, $\hat{V}$ the specific volume, R the gas constant, and the remaining parameters defined as

$$a = 0.42747 \frac{(RT_c)^2}{P_c}$$

$$b = 0.08664 \frac{RT_c}{P_c}$$

$$m = 0.48508 + 1.55171\omega - 0.1561\omega^2$$

$$\alpha = \left[1 + m\left(1 - \sqrt{T/T_c}\right)\right]^2,$$

where T_c and P_c are the critical temperature and pressure, respectively, for the substance of interest and ω the acentric factor for the substance.

Equations of state like the SRK equation relate T, P, and $\hat{V}$ for a given substance, and whenever two of the three parameters are known, the EOS can be used to determine the third, unknown parameter. If T and $\hat{V}$ are given for a known substance, then it is trivial to calculate P after looking up T_c, P_c, and ω for that substance. However, the most common situation is that T and P are known, and we wish to calculate $\hat{V}$. This is a much more difficult problem when the SRK equation of state is used because it requires the solution to a cubic equation.

Chemical and Biomedical Engineering Calculations Using Python®, First Edition. Jeffrey J. Heys.
© 2017 John Wiley & Sons, Inc. Published 2017 by John Wiley & Sons, Inc.
Companion Website: www.wiley.com/go/heys/engineeringcalculations_python

Our goal in the first half of this chapter is to examine two different methods for finding solutions to nonlinear, algebraic equations like the SRK equation when $\hat{V}$ is unknown. The first method, bisection, is slow but very robust, while the second method, Newton's method, is fast but does not always converge to the solution.

Before examining bisection and Newton's method in detail, we can use some functions included with the Scipy library to solve the SRK problem described above. It is, of course, dangerous to use an algorithm that we do not understand – so consider this approach with skepticism, but having a solution to our example problem will be helpful going forward. The Scipy library includes a number of functions for solving nonlinear equations, and these functions are contained in the optimization section of the library. Some of the nonlinear solvers are designed for a single nonlinear equations and other solvers, examined later in this chapter, are designed for large systems of nonlinear equations. Here, we will use the "broyden1" function (`scipy.optimize.broyden1()`) to solve the SRK equation for $\hat{V}$. The use of the broyden1 function (and all other nonlinear solvers) requires at least two inputs: (1) the name of a function containing the nonlinear equation and (2) a guess for the solution. Further, the nonlinear function must be rearranged so that all the terms are on one side of the equal sign and it returns zero at the solution. Hence, if we have a general nonlinear equation like the SRK equation that can be written as $f(x) = g(x)$ and we are searching for x that satisfies this equality, we must rewrite the function as $f(x) - g(x) = 0$ so that the function returns zero when x is found.

Nonlinear Solution Algorithms

For most algorithms that find solutions to a nonlinear equation, the nonlinear equation must be rearranged so that it is equal to zero at the solution.

The SRK equation should be rewritten as

$$P - \frac{RT}{\hat{V} - b} + \frac{\alpha a}{\hat{V}(\hat{V} + b)} = 0. \tag{8.2}$$

The Python script below begins with the definition of the SRK function, which is passed a guess for $\hat{V}$ as the only input, and then using that guess and the appropriate constants for the substance of interest (carbon monoxide in this case), tests to see if the SRK equation is satisfied. If the function returns zero, then the estimated value for $\hat{V}$ was correct and the original function evaluates to zero, that is, it is a root of the function.

```
import math
import scipy.optimize
```

```
def SRK(V):
    # Properties of Carbon Monoxide
    T = 300 # K
    P = 10 # atm
    Tc = 133.0 # K
    Pc = 34.5 # atm
    w = 0.049 # Acentric factor for CO
    R = 0.08206 # L atm / (mol K)
    a = 0.42747*(R*Tc)**2 / (Pc)
    b = 0.08664*(R*Tc/Pc)
    m = 0.48508+1.55171*w-0.1561*w**2
    alpha = (1+m*(1-math.sqrt(T/Tc)))**2
    term1 = R*T/(V-b)
    term2 = alpha*a/(V*(V+b))
    return P-term1+term2


V = 2.0 # L/mol, initial guess

V = scipy.optimize.broyden1(SRK, V, maxiter=100, f_tol=1e-6)
print(V)
```

The Python script above should print out 2.46 L/mol as a root to the SRK equation, and this is the desired answer. At this point, we do not know how the broyden1() function determines this root, but, it turns out that this function is similar to Newton's method, which is covered later in this chapter.

Finally, it was noted previously that the SRK equation is a cubic equation in terms of $\hat{V}$, which implies that there could be up to three possible roots. One method for trying to find the other possible solutions is to change the initial guess for $\hat{V}$. Trying this approach reveals that the broyden1 function converges to 2.46 L/mol for any strictly positive guess. For any strictly negative guess, the method fails to converge to a solution, and a guess of 0.0 gives NaNs ("Not a Number") errors. These errors are usually caused by dividing by zero. This reinforces the issue of robustness with nonlinear solvers. In this simple case of one equation, the function can be plotted to show that there is likely only one root. This plot will be revisited to address the issue of robustness.

8.2 Bisection Method

The bisection method is based on identifying a region for the independent variable that bounds the root that we are trying to find. In Figure 8.1, a nonlinear function crosses the x-axis at the solution and points $x = a$ and $x = b$ have been identified that bound the solution. The bisection algorithm is based on an iteration with the following steps:

1) calculate the midpoint of the region bounding the root, $c = (a + b)/2$,
2) calculate the value of the function, $f(c)$ at the midpoint, and

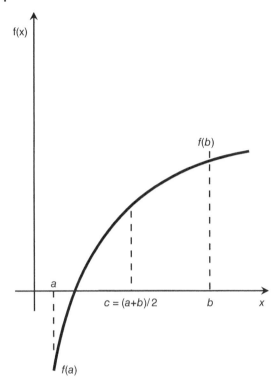

Figure 8.1 The bisection method is used to find the roots or zeros of a nonlinear function. The first step is to identify points a and b such that $f(a) \cdot f(b) < 0$ (i.e., $f(a)$ and $f(b)$ have different signs). The distance between a and b is halved and then a or b is discarded depending on the sign of $f(c)$. This process is repeated until the location of the root within some tolerance is found.

3) determine which previous endpoint, $f(a)$ or $f(b)$, has the same sign (positive or negative) as $f(c)$ and replace that endpoint with c.

In Figure 8.1, endpoint b would be replaced by c because $f(b)$ and $f(c)$ have the same sign (positive), and the new region that bounds the root would be between a and c. Notice that the region bounding the endpoint is halved every iteration and, eventually, $f(c)$ is close to zero and the magnitude of $f(c)$ is less than a preset tolerance. The bisection method is extremely robust in that once an a and b that bound the root are known, it *always* converges to a solution (i.e., a root of the equation). The weaknesses of this approach are the requirement that a and b be determined manually (fortunately for many problems, we have a rough estimate for the solution) and convergence is slow because we are only halving the domain each iteration.

The Python script below uses the bisection method to find a root, that is, find a value for $\hat{V}$ that satisfies the SRK equation.

```python
import math

def SRK(V):
    # Properties of Carbon Monoxide
    T = 300 # K
```

```python
    P = 10 # atm
    Tc = 133.0 # K
    Pc = 34.5 # atm
    w = 0.049 # Acentric factor for CO
    R = 0.08206 # L atm / (mol K)
    a = 0.42747*(R*Tc)**2 / (Pc)
    b = 0.08664*(R*Tc/Pc)
    m = 0.48508+1.55171*w-0.1561*w**2
    alpha = (1+m*(1-math.sqrt(T/Tc)))**2
    term1 = R*T/(V-b)
    term2 = alpha*a/(V*(V+b))
    return P-term1+term2


#Settings
maxIter = 1000
TOL = 1e-4
stop = False

# Bisection method setup
a = 1.0
fa = SRK(a)
b = 5.0
fb = SRK(b)

# Check starting values
if(fa*fb >= 0):
    print("ERROR: bad starting bounds")
else:
    # Bisection iteration
    for i in range(maxIter):
        c = (a+b)/2
        fc = SRK(c)
        print("c = ", c, " and f(c)= ", fc)
        # Check for convergence
        if(math.fabs(fc) < TOL):
            print("Root found at ", c)
            stop = True
            break

        # if not converged, determine end point
        # to replace
        if(fa*fc < 0):
            b = c
            fb = fc
        else:
            a = c
            fa = fc
```

```
if stop == False:
    print("Reached maxIter. Currently at ", c)
```

When this script is run, it prints out the midpoint, c, and $f(c)$ every iteration. Depending on the initial bounds, 15–20 iterations are required for convergence to the tolerance of 1×10^{-4}. With nonideal gas problems like this, it is possible to obtain a good initial estimate for the solution using the ideal gas law, $P\hat{V} = RT$. For carbon monoxide under the conditions listed above, the ideal gas law gives $\hat{V} = 2.46$ L/mol, which is almost identical to the solution obtained using the SRK equation. If the pressure increases or the temperature decreases, the ideal gas law becomes less accurate.

Debugging Practice

A novice programmer rewrote a few lines of the bisection script as shown below:

```
# if not converged, determine end point
# to replace
if(fa*fc < 0):
    c = b
    fc = fb
else:
    c = a
    fc = fa
```

The resulting method failed to converge. Why?

8.3 Newton's Method

The most useful equation in computational mathematics may be the Taylor polynomial (also known as Taylor series or Taylor expansion). Newton's method is one of many results given in this book that may be derived using a Taylor polynomial. A Taylor polynomial is an expansion of $f(x)$ about a nearby point, x_0:

$$f(x) = f(x_0) + (x - x_0)f'(x_0) + \frac{(x - x_0)^2}{2!}f''(x_0) + O(x - x_0)^3. \tag{8.3}$$

Typically, x_0 is chosen to be a point where information about the function $f()$ is known. The last term tells us that the magnitude of the term is on the order of $(x - x_0)^3$, but the exact term is not given. If the distance between x and x_0 is small (i.e., $x - x_0 \ll 1$), then $(x - x_0)^3$ or even $(x - x_0)^2$ is very small.

In Newton's method, we are trying to find x such that $f(x) = 0$. Assuming that we have a guess, called x_0, for the value of x and setting $f(x) = 0$, the

Taylor polynomial, ignoring the last two terms given above because they are (hopefully) small, becomes

$$0 = f(x_0) + (x - x_0)f'(x_0).$$

Rearranging and solving for the unknown x gives

$$x = x_0 - \frac{f(x_0)}{f'(x_0)}. \tag{8.4}$$

Figure 8.2 illustrates how this equation helps us to find the root of interest. Noticing that we only retained the linear term from the Taylor polynomial, Newton's method approximates the nonlinear function with a line at current estimate for the location of the root, x_0, and that line is then used to calculate a new and hopefully better estimate for the location of the root. A new Taylor polynomial is used to approximate the nonlinear function with a line at this new location, and the process is repeat until the root is found (in the case of convergence) or an iteration limit is reached (failure).

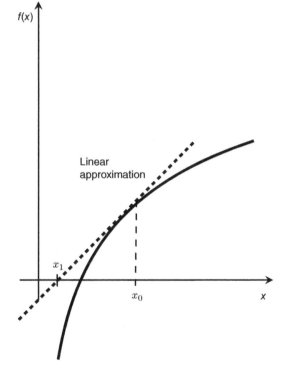

Figure 8.2 Newton's method approximates a nonlinear function with a straight line at the point x_0. The linear approximation is then used to obtain a new estimate for the root (i.e., the intersection of the nonlinear function with the x-axis) of the nonlinear function.

Newton's method is used to solve for the roots of the equation $f(x) = \sin(x) - x^2$ in the Python script below.

```
import math

TOL = 1.0e-6
MAX_ITER = 10

def f(x): # function
    return numpy.sin(x) - x*x

def df(x): # derivative
    return math.cos(x) - 2.0*x

itr = 0    # iteration counter
x = 1.0    # initial guess at root
res = f(x) # initial function evaluation

while abs(res) > TOL and itr < MAX_ITER:
    itr += 1
    res = f(x)
    print('Iteration: ', itr, 'res: ', res)
    x = x - res/df(x)

if abs(res) < TOL:
    print('Converged to ', x)
else:
    print('Did not converge')
```

The script uses a **while** loop with two conditions: (1) the value of $f(x)$ at the current estimate for x must remain larger than the convergence tolerance and the iteration counter must be less than the maximum iterations allowed. Upon execution, Newton's method converges to a root a 0.877 in four iterations. Figure 8.3 is a plot of $f(x)$ for $x \in [0.0, 1.5]$ and confirms that there is a root at the location that Newton's method converged. There is a second root at $x = 0.0$, and this root can be found by changing the initial guess to $x = 0.1$ or some other value less than 0.5.

Newton's method has two advantages over the bisection method:

1) Only a single estimate for the location of the root is required.
2) The error in approximating the nonlinear function with a straight line is on the order of $(x - x_0)^2$. As the guess, x_0, gets close to the true solution, x, the linear approximation is very good and converge is extremely fast (i.e., convergence is quadratic near the solution).

Figure 8.3 The nonlinear function $f(x) = \sin(x) - x^2$ with roots at $x = 0.0$ and $x = 0.877$.

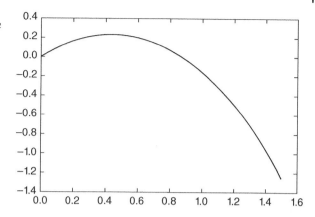

Newton's method also has two disadvantages compared to the bisection method:

1) We need to have a function that calculates the derivative, $f'(x)$, at a point, x_0. This limitation, however, is somewhat overcome in the next section on Broyden's method.
2) If the initial guess for the location of the root is not sufficiently accurate, the iterations can diverge and a solution will not be found. This behavior was seen previously using Broyden's method when the initial estimate of $\hat{V}$ was less than zero.

8.4 Broyden's Method

One disadvantage associated with Newton's method is the requirement that the derivative of the function of interest also be provided to the algorithm. Having an approximation for the derivative would be a significant benefit if it were available. Fortunately, the Taylor polynomial can once again help us by providing an approximation to the derivative. Recall the Taylor polynomial:

$$f(x) = f(x_0) + (x - x_0)f'(x_0) + \frac{(x - x_0)^2}{2}f''(x_0) + O(x - x_0)^3, \tag{8.5}$$

and notice that if the higher order terms are discarded (i.e., the last two terms on the right side), it can be rearranged to provide an estimate for the derivative:

$$f'(x_0) \approx \frac{f(x) - f(x_0)}{(x - x_0)}.$$

If we substitute this directly into the Newton iteration, everything will cancel out because we just used the same equation twice. Instead of using the current guess, x_0, and the next guess, x, to approximate the derivative, we can use the previous two guesses:

$$f'(x_1) \approx \frac{f(x_1) - f(x_0)}{x_1 - x_0}. \tag{8.6}$$

Notice that the current guess is now x_1 and the previous guess was x_0. Substituting this approximation for the derivative into Newton's method gives

$$x_{new} = x_1 - \frac{f(x_1)(x_1 - x_0)}{f(x_1) - f(x_0)}. \tag{8.7}$$

This approach is also frequently called the Secant method because the nonlinear function is approximated using a secant line instead of a tangent line.

The Python script below uses Broyden's method (equation 8.7) to find the solution to the SRK model problem that was solved previously.

```python
import math

def SRK(V):
    # Properties of Carbon Monoxide
    T = 300 # K
    P = 10 # atm
    Tc = 133.0 # K
    Pc = 34.5 # atm
    w = 0.049 # Acentric factor for CO
    R = 0.08206 # L atm / (mol K)
    a = 0.42747*(R*Tc)**2 / (Pc)
    b = 0.08664*(R*Tc/Pc)
    m = 0.48508+1.55171*w-0.1561*w**2
    alpha = (1+m*(1-math.sqrt(T/Tc)))**2
    term1 = R*T/(V-b)
    term2 = alpha*a/(V*(V+b))
    return P-term1+term2

#Settings
maxIter = 1000
TOL = 1e-4

# Broyden Setup
x0 = 1.0
fx0 = SRK(x0)
x1 = x0+1e-2
fx1 = SRK(x1)

for i in range(maxIter):
    xNew = x1 - fx1*(x1-x0)/(fx1-fx0)
    fxNew = SRK(xNew)
```

```
print("xNew = ", xNew, " and fxNew = ", fxNew)
if(math.fabs(fxNew) < TOL):
    print("Root found at ", xNew)
    break
else:
    x0 = x1
    fx0 = fx1
    x1 = xNew
    fx1 = fxNew

if(math.fabs(fxNew) > TOL):
    print("Reached maxIter. Current estimate at ", xNew)
```

Using Broyden's method, the solution, $\hat{V} = 2.46$, is obtained in about seven iterations, depending on the initial guess. The iterations are especially rapid near the solution with $f(x) = -0.00016$ after six iterations and $f(x) = 1.7 \times 10^{-7}$ after seven iterations. Notice that the value of $f(x)$ is converging to zero quadratically (or nearly quadratically since we are approximating the derivative.

Using an initial guess of $x_0 = -1.0$ once again leads to an error message (in this case division by zero) and failure to converge. Now that we understand the principle behind Newton's method and Broyden's method, we can examine why this occurs. Figure 8.4 shows the nonlinear SRK function with a single root at $\hat{V} = 2.46$. For any guess near that root, if we approximate the SRK function with a straight line and then notice where that line cross the x-axis, it will probably

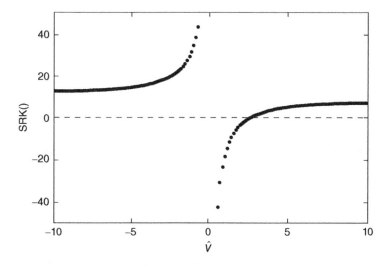

Figure 8.4 The nonlinear SRK function with a single root at $\hat{V} = 2.46$.

be a point closer to the solution, and the iterations will converge. If our initial guess is negative, however, any linear approximation to the function is going to predict that the root lies further to the left. As a result, each iteration will take us further and further toward $x \to -\infty$.

8.5 Multiple Nonlinear Equations

Problems that have multiple nonlinear equations that must all be solved simultaneously arise in problems as diverse as fluid piping networks (i.e., water distribution systems) to staged distillation columns. In order to solve these systems of nonlinear equations, we need to use the tools we have already learned for linear systems (Chapter 6) and single nonlinear equations (this chapter) and combine them together. The first step is to recall Newton's method:

$$x_{new} = x_0 - \frac{f(x_0)}{f'(x_0)}, \tag{8.8}$$

which enables the iterative determination of x such that $f(x) = 0$. Now we are interested in the case where we have n nonlinear equations (some prefer to think of this as a *vector* of equations) written $\mathbf{f}(\mathbf{x}) = \mathbf{0}$ and we are trying to determine the vector of unknowns, $\mathbf{x}$, that satisfies this system of equations. For multiple equations, Newton's method can be rewritten:

$$\mathbf{x_{new}} = \mathbf{x_{old}} - \frac{\mathbf{f}(\mathbf{x_{old}})}{\mathbf{f'}(\mathbf{x_{old}})}, \tag{8.9}$$

where

$$\mathbf{f'}(\mathbf{x}) = \begin{bmatrix} \dfrac{\partial f_0}{\partial x_0} & \dfrac{\partial f_0}{\partial x_1} & \cdots \\ \dfrac{\partial f_1}{\partial x_0} & \dfrac{\partial f_1}{\partial x_1} & \cdots \\ \vdots & & \ddots \end{bmatrix} \tag{8.10}$$

and

$$\mathbf{f}(\mathbf{x}) = \begin{bmatrix} f_0(\mathbf{x}) \\ f_1(\mathbf{x}) \\ \vdots \end{bmatrix}. \tag{8.11}$$

All of the previous advantages (fast, single initial guess) and disadvantages (calculate derivatives, may not converge) apply when Newton's method is applied to a system of nonlinear equations instead of a single equation. It is common to call $\mathbf{f}(\mathbf{x})$ the *residual* or residual vector and $\mathbf{f'}(\mathbf{x})$ is the *Jacobian* or Jacobian matrix. Calculating the residual divided by the Jacobian requires solving a linear matrix problem: $\mathbf{f'}(\mathbf{x}) \cdot \mathbf{dx} = \mathbf{f}(\mathbf{x})$ for $\mathbf{dx}$. Once $\mathbf{dx}$ is determined, then

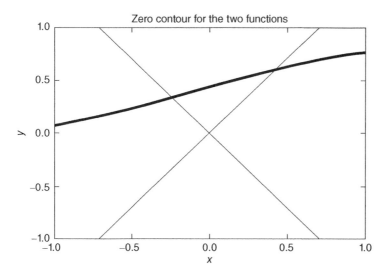

Figure 8.5 A plot of the zero contour lines (i.e., the lines where f_0 and f_1 equal zero) for the example functions. The thicker contour line is f_1. The two points where both functions are zero are solutions to this nonlinear system, and two different solution vectors are seen to exist in this plot.

$\mathbf{x_{new}} = \mathbf{x_{old}} - \mathbf{dx}$. This is wrapped in an iteration and repeated until $\|\mathbf{f(x)}\|_2 <$ *TOL*. In other words, if $\mathbf{x}$ is close to the solution vector, then each equation in the residual should be a small number and the L^2-norm of the residual should be small.

As an example, consider the system of two nonlinear equations with two unknowns:

$$f_0(x, y) = 2 * x^2 - y^2,$$
$$f_1(x, y) = y - 0.5 \cdot (\sin(x) + \cos(y)).$$

For a system this small, it is possible to use a contour plot to explore the properties of the solution. Putting both of the unknowns, x and y, into a single vector: $x[0] = x$ and $x[1] = y$ and plotting the contours where each function is equal to zero generates the plot shown in Figure 8.5.

A Python script that uses Newton's method to solve this small system of two nonlinear equations is shown below.

```python
import numpy
import numpy.linalg as nl
import math

def res(x):
    n = x.size
```

```
    f = numpy.zeros(n,dtype=numpy.float64)
    f[0] = 2*x[0]**2 - x[1]**2
    f[1] = x[1]-0.5*(math.sin(x[0])+math.cos(x[1]))
    return f

def jac(x):
    n=x.size
    j = numpy.zeros((n,n),dtype=numpy.float64)
    j[0,0] = 4*x[0]
    j[0,1] = -2*x[1]
    j[1,0] = -0.5*math.cos(x[0])
    j[1,1] = 1.0 + 0.5*math.sin(x[1])
    return j

x = numpy.array([1.0,1.0])
TOL = 1.0e-6
maxIter = 20
for i in range(maxIter):
    f = res(x)
    j = jac(x)
    print("Iteration ",i," has a norm of ",nl.norm(f,2))
    if (nl.norm(f,2) < TOL):
        print("Converged in ", i, " iterations to ", x)
        break
    else:
        dx = nl.solve(j,f)
        x = x - dx
if(i == maxIter):
    print("Failed to converge in ", maxIter, " iterations")
```

The residual vector and Jacobian matrix are calculated in separate functions that are called each Newton iteration. The L^2-norm of the residual vector is then calculated to check for convergence. If convergence has not been achieved, then the linear matrix problem is solved and the guess at the solution is updated based on that solution. Using an initial guess of $(1.0, 1.0)$, Newton's method required five iterations to converge to one of the possible solutions, $[0.44, 0.62]$. It also displayed rapid, quadratic convergence once the approximate solution was close to the final solution.

As with Broyden's method, it is possible to avoid having to calculate an exact Jacobian matrix. Numerical approximations for all the different derivatives within the Jacobian matrix can be used instead of writing an explicit Jacobian function. Most of the algorithms in the `scipy.optimize` library for solving systems of nonlinear equations are capable of calculating an approximate Jacobian matrix automatically. The Python script below demonstrates three different algorithms available in `scipy.optimize` for solving linear systems of equations.

```
import numpy
import math
from scipy.optimize import newton_krylov

def res(x):
    n = x.size
    f = numpy.zeros(n,dtype=numpy.float64)
    f[0] = 2*x[0]**2 - x[1]**2
    f[1] = x[1]-0.5*(math.sin(x[0])+math.cos(x[1]))
    return f

x0 = numpy.array([0.0,0.0])
x = fsolve(res,x0)
print("Converged to ", x)
x = newton_krylov(res, x0)
print("Converged to ", x)
x = broyden1(res, x0)
print("Converged to ", x)
```

Only the original nonlinear functions are provided to these algorithms, and the algorithms automatically approximate the Jacobian. All three of these algorithms, `fsolve()`, `broyden1()`, **and** `newton_krylov()` are basically Broyden's method, which was discussed previously, extended to multiple equations. Interestingly, using an initial guess of $(0.0, 0.0)$, two of the methods, `fsolve()` **and** `broyden1()`, converged to the solution at $x = (0.44, 0.62)$, but `newton_krylov()` converged to the other solution, $x = (-0.25, 0.35)$.

8.5.1 The Point Inside a Square

Imagine that you are asked to solve the following geometric puzzle: a square with sides of unknown length contains a point that is exactly 3, 4, and 6 ft from successive corners of the square. What is the length, in feet, of the side of the square? You begin by drawing the diagram shown in Figure 8.6.

Using Figure 8.6, equations based on the Pythagorean theorem can be derived. Starting with the upper-right triangle that has a hypotenuse of length 3, and using the edge lengths labeled a, b, c, and d, the Pythagorean theorem states that

$$a^2 + c^2 = 3^2 = 9.$$

For the lower left triangle, the Pythagorean theorem states that

$$b^2 + c^2 = 16,$$

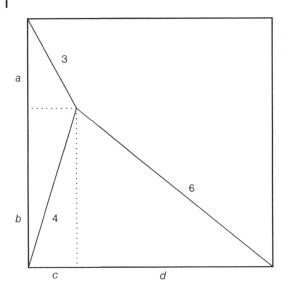

Figure 8.6 Sketch of a square containing a point that is 3, 4, and 6 ft from successive corners.

and for the triangle in the lower right, the Pythagorean theorem states that

$$b^2 + d^2 = 36.$$

At this point, we have four unknowns (*a*, *b*, *c*, and *d*) and three nonlinear equations containing various combinations of those unknowns. In order to have a potentially solvable system of equations, we need one additional equation relating the unknowns. This final equation can be obtained by noting that each side of the square must have the same length, so

$$a + b = c + d.$$

Now we have a nonlinear system of four equations with four unknowns. Before using Broyden's method to solve for the four unknowns, it is necessary to rewrite the equations so that all the terms in every equation are on one side of the equal sign.

The Python script below uses `scipy.optimize.broyden1()` to approximate the solution to this system of equations.

```
import scipy.optimize

def equations(vals):
    a, b, c, d = vals
    eq1 = a + b - c - d
    eq2 = b**2 + d**2 - 36
```

```
eq3 = a**2 + c**2 - 9
eq4 = c**2 + b**2 - 16
return (eq1, eq2, eq3, eq4)

sol = scipy.optimize.broyden1(equations,(3,3,3,3))
print(sol)
print(sol[0]+sol[1])
print(sol[2]+sol[3])
```

The script begins by importing the `scipy.optimize` library, and then it has a function containing the four equations being solved. The only input into this function is a vector containing estimated values for the four unknowns, and this vector is unpacked so that the estimated values are stored in the variables a, b, c, d and are easier to use in the four equations. The script ends by printing the solution found by Broyden's method and printing the length of the two sides of the square. The two sides should, of course, be equal to each other. For this particular point to be 3, 4, and 6 ft from successive corners of a square, the square must have sides that are approximately 6.27 ft.

Problems

8.1 Use Newton's method to find a solution for $\sin(x) - x = 0$. This equation is very similar to the equation solved in the example problem in Section 8.3, so modifying that algorithm may be the simplest approach. Compare the convergence of Newton's method on this problem ($\sin(x) - x = 0$) to the convergence of Newton's method on the example problem ($\sin(x) - x^2 = 0$). Why is the convergence behavior different? Plot both functions because that may provide some insight. Hint: calculate each term in the Newton's method equation at the solution.

8.2 Use Newton's method to find a solution for $x^3 + 2x^2 - 2 = 0$. This function displays some interesting behavior when using Newton's method, depending on the initial guess. There is only a single solution (near $x = 0.8$), but an initial guess between $x \in (-2, 0)$ is unlikely to converge to the one solution while other initial guesses will converge to the solution. Explain this behavior (hint: plot the function).

8.3 You have been hired by Con Edison because there is a giant Arctic cold front approaching New York, and they are worried about condensation of natural gas. We will assume that natural gas is a mixture of methane and ethane. The following equations will allow you to estimate the dew point temperature (i.e., the temperature where condensation initiates) for natural gas.

The dew point temperature is the temperature at which the following equations is satisfied:

$$750 \text{ mm Hg} = y_M P_M^* + y_E P_M^*,$$

where y_M is the mole fraction of methane in the natural gas, y_E the mole fraction of ethane in the natural gas, and P^* the vapor pressure of methane or ethane, depending on the subscript. The abovementioned equation assumes that atmospheric pressure (or barometric pressure) in New York is 750 mm Hg. The vapor pressures of methane and ethane depend on the temperature, and they are calculated using Antoine's equation:

$$P^* = 10^{\left(A - \frac{B}{T+C}\right)},$$

where T is the temperature in °C, P^* the vapor pressure in mm Hg, and A, B, and C the species-specific constants given in the table below. Note that the term after the number "10" is an exponent.

	Methane	Ethane
A	6.61184	6.80266
B	389.93	656.4
C	266.0	256.0

Con Edison typically has natural gas that has $y_M = 0.90$ (i.e., 90 mol% methane) and $y_E = 0.10$ (i.e., 10 mol% ethane). Determine the dew point temperature by solving the nonlinear equation above. Con Edison would also like to know how the dew point temperature changes if the composition of the natural gas changes to 70 mol% methane and 30 mol% ethane. Should they be worried about condensation occurring in New York for the "worst case scenario" of Arctic cold front?

8.4 A new large backyard gas fire pit has recently been developed that is fueled by propane. Unfortunately, the propane tank was painted black so that it would be easier to hide in a backyard, but the black color causes the tank contents to heat up in the hot sun and some tanks have ruptured. You have been hired by the manufacturer to calculate the quantity of propane in the tank at the time of rupture. The manufacturers believe that van der Waal's equation of state is an accurate equation for these conditions. The equation is

$$\left(P + \frac{a}{\hat{V}^2}\right)(\hat{V} - b) - RT = 0,$$

where

$$a = \left(\frac{27}{64}\right)\left(\frac{R^2 T_c^2}{P_c}\right),$$

$$b = \left(\frac{1}{8}\right)\left(\frac{RT_c}{P_c}\right).$$

The following properties for propane and other constant were provided to you:

- $T = 384$ K
- $P = 4891.3$ kPa
- tank volume is 0.15 m³ so $\hat{V} = \frac{0.15\,\text{m}^3}{n}$ where n is moles of propane in the tank
- $T_c = 369.9$ K (critical temperature of propane)
- $P_c = 4254.6$ kPa (critical pressure of propane)
- gas constant: $R = 0.008314$ m³ kPa/(mol · K)

The pressure and temperature provided to you are the last pressure and temperature readings on the tank just before rupture. Solve van der Waal's equation for the number of moles in the 0.15 m³ tank at the time of rupture. Note that the temperature and pressure both exceed the critical values so the fluid in the tank is not a liquid or a gas but is a super critical fluid. You should report the quantity of super critical propane in the tank in terms of moles and mass (kg). You should also generate a plot of the value of the left side of van der Waal's equation for different values of n, the number of moles in the tank, to convince the manufacture that there is only a single physically possible solution to the equation.

8.5 The trajectory of any projectile object (under certain assumptions) can be determined using Newton's laws of motion, $F = m \cdot a$. After a football has been released by a quarterback, the primary force acting on the football is gravity, thus, the acceleration, a, of the football can be described by

$$0 = m \cdot a_x, -m \cdot g = m \cdot a_y,$$

where a_x is acceleration in the x- or horizontal direction, a_y the acceleration in the y- or vertical direction, m the mass of the football, and $g = 9.806$ m/s². Since acceleration, a, equals the derivative of velocity with respect to time, $\frac{dv}{dt}$, the acceleration equations can be integrated to give equations for velocity:

$$v_x = v_0 \cos(\theta)$$

$$v_y = v_0 \sin(\theta) - g \cdot t,$$

where v_0 is the initial velocity of the football and θ is the initial upward angle of the throw. Since velocity, v, equals the derivative of location, (x, y), with respect to time, the velocity equations can be integrated to give equations for location:

$$x = x_0 + v_0 \cos(\theta) \cdot t$$
$$y = y_0 + v_0 \sin(\theta) \cdot t - \frac{1}{2} \cdot g \cdot t^2,$$

where (x_0, y_0) is the initial location of the quarterback. Solving the x location equation for time gives

$$t = (x - x_0)/(v_0 \cos(\theta))$$

and substituting this equation for time into the y location equation (and moving everything to one side of the equal sign) gives

$$0 = y_0 + \tan(\theta) \cdot (x - x_0) - \frac{g \cdot (x - x_0)^2}{2 \cdot v_0^2 \cdot \cos^2(\theta)} - y.$$

You have been hired by the quarterback of the local professional (American) football team to determine the angle, θ, for throwing the football so that it travels downfield for 50 m and can be caught 2 m above the ground. Use the following data (obtain from the NFL Scouting Combine) in your analysis:

- Quarterback location (x_0, y_0) is $(0.0, 0.0)$
- Quarterback arm strength gives an initial velocity of 25.0 m/s

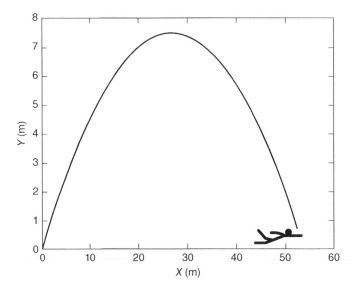

Figure 8.7 Trajectory of the football in the (x, y) plane.

- Target location (x, y) is $(50.0, 2.0)$ in m
- Broyden's method as implemented in Scipy is recommended.

Your final report to the quarterback should include the initial angle, θ for the throw in radians and degrees, and you should have a plot of the ball location, (x, y) over time (Figure 8.7). Use this plot to estimate the time to target to the nearest second.

9

Statistics

9.1 Introduction

The rigorous statistical analysis of data is a critical aspect of our never ending pursuit of *more* – more efficiency, more consistency, more profitability, and so on. There is a large number of outstanding software packages that facilitate the processes of statistical analysis. Commercial packages such as IBM SPSS Statistics and SAS are widely used, and open source (free) packages such as R are well-documented and have large communities of users. The motivation for *briefly* examining select statistical functions available in the `scipy.stats` library is that it is often helpful to combine statistical analysis with other mathematical computations. For example, we might be using a Python script to model a fermentation reactor, and one input to that model is the dissolved oxygen (DO) concentration for the inflow stream. Before running the reactor simulation, we might want to calculate the averaged DO concentration as well as the standard deviation. Being able to perform all these calculations – from solving ordinary differential equations to standard deviation calculations – in one software package or programming environment can help us to be more efficient and make fewer mistakes. Transferring results from one computer to another computer or one software platform to another software platform is a time-consuming process that frequently introduces mistakes. Thus, it should be avoided whenever possible.

9.2 Reading Data from a File

Before data can be statistically analyzed, it often needs to be loaded from a file. In this section, a few different options are examined for loading data from a comma separated value or csv file. This format is used here because it is a common format for sharing data between different software packages. Most spreadsheets, including Excel, can export the contents of a spreadsheet as a csv file. The goal of the different methods presented in this section is to import the

Chemical and Biomedical Engineering Calculations Using Python®, First Edition. Jeffrey J. Heys.
© 2017 John Wiley & Sons, Inc. Published 2017 by John Wiley & Sons, Inc.
Companion Website: www.wiley.com/go/heys/engineeringcalculations_python

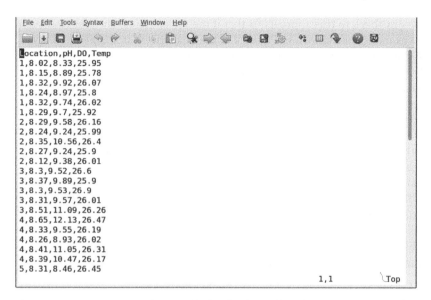

Figure 9.1 A screen capture showing the first few lines of the file "DOdata.csv". The file contains a header row followed by the data.

contents of a csv text file into a numpy array for later statistical analysis and plotting.

The first data set that will be imported contains multiple pH, dissolved oxygen (DO), and temperature data from six different locations in a river. The locations are distinguished only by a number (1 through 6). Figure 9.1 shows a screen capture of the contents of the first part of the data file (DOdata.csv). The file contains a header row that contains the names of the various columns of data and the actual data follows below.

9.2.1 Numpy Library

The simplest method for importing data from a file into a numpy array is to use the function numpy.genfromtxt(). This function is passed as inputs: (1) the name of the file; (2) the delimiter (i.e., the character separating the unique data points), which is typically a comma or a tab character; and (3) the number of header rows to skip. The numpy documentation includes descriptions for other, optional inputs for the genfromtext() function. The Python script below illustrates the use of this function for importing data from a csv file.

```
import numpy

data = numpy.genfromtxt("DOdataNoHeader.csv", \
  delimiter=",",skip_header=1)
numRows, numCols = data.shape
```

One drawback to this approach is that if the file contains a header row, such as the example data shown in Figure 9.1, the header text is skipped and is not readily available for use later in the code. Another drawback is that all the data in the file should be the same data type. If one column contains, for example, text data or missing data, the import function returns an error. If this approach is used, the csv file should be carefully checked that it only contains numerical data and any header text should be recorded separately if it is needed.

A second approach for importing data from a file using the numpy library starts by using some standard Python functions for opening a file and then reading the file contents one row at a time. Every time a new row is read in from the file, it is loaded into the numpy array using the numpy.fromstring() function. This approach requires the construction of an empty numpy array to hold new data as it is read in from the file. The advantage of this approach is that it allows one to load the header information from the file and store it in a separate variable from the numpy array. The Python script below implements this more manual approach for reading in data from a cvs file.

```python
import numpy

# open file containing data
dataFile = open("DOdata.csv")

# read first line with headers and split them up
header = dataFile.readline()
dataHeaders = header.split(',')
numCols = len(dataHeaders)

# set maxRows we are capable of reading
maxRows = 256
data=numpy.zeros((maxRows,numCols),
                 dtype=numpy.float64)

# read in data one row at a time
rowCount = 0
for row in dataFile:
    data[rowCount,:] = numpy.fromstring(row,sep=',')
    rowCount += 1
    if(rowCount >= maxRows):
        break
```

A few important observations can be made from this routine. First, both the split() function and the numpy.fromstring() function are passed a character for delimiting between columns. Second, a parameter limiting the max number of rows of data that can be loaded from the file is set. This is recommended for security (avoids accidentally reading in 10 GB worth of data) and computational speed (avoids have to resize the numpy array every time new data is loaded).

9.2.2 CVS Library

Another approach for loading data from a cvs file takes advantage of the cvs library that is a standard part of the Python language. This approach still allows special handling of the header row, and it also requires the creation of an empty numpy array for storing the data as it is read from the file. When each row is read from the file, it is imported as a list of strings, that is, each number is a unique string. These strings are converted into floating point numbers using the function numpy.astype(**float**). Once all the rows are read from the file, the numpy array can be resized to truncate any empty rows from the end of the array that was originally created. The Python script below implements this third approach.

```python
import csv
import numpy

with open('DOdata.csv', 'r') as dataFile:
    reader = csv.reader(dataFile)
    header=next(reader)
    numCols = len(header)
    maxRows = 256
    data = numpy.zeros((maxRows,numCols))
    rowCount = 0
    for row in reader:
        data[rowCount,:]=numpy.array(row).astype(float)
        rowCount += 1
        if(rowCount >= maxRows):
            break
    data = numpy.resize(data,(rowCount,numCols))
```

9.2.3 Pandas

The final method for importing data from a file is to use the pandas library (pandas.pydata.org). The pandas library was developed from the very beginning to enable the analysis of large data sets in the most simple and straightforward manner possible. As a result, pandas includes functions for importing data from a wide variety of file types including csv files, Excel files, and many others. Pandas can handle heterogeneous data sets that include columns of floating point numbers next to columns of string data. Further, pandas can import data sets with missing data, and it includes functions that can attempt to fill-in the missing data using a number of standard approaches if that is necessary. The pandas library is one of the most actively developed libraries for data analysis using Python, and further exploration of the library and its features is highly encouraged for anyone that routinely performs data analysis tasks [1].

The use of pandas to import the DO data set used in this chapter is explored through a brief example. Pandas has a number of built-in functions for importing and exporting data sets to files including `read_csv()`, `read_excel()`, `read_html()`, and `to_csv()`. The full description of input/output functions can be found on the pandas website (http://pandas.pydata.org/pandas-docs/stable/io.html). The Python script below illustrates the `read_csv()` function for importing the DO data set:

```python
import numpy
import pandas
import matplotlib.pyplot as plt

data = pandas.read_csv("../../data/DOdata.csv")
```

This is remarkably simple when compared to the other methods we have examined for importing data. The function automatically recognizes the header information and it automatically recognizes the use of commas to separate the data. The information in the file is stored in the variable "data", which is technically a pandas Data Frame object. As expected, it is relatively simple to access the data within the Data Frame. The code below is a continuation of the previous Python script, and it illustrates how to access, modify, and analyze the data that was read from the file.

```python
print(data.columns)
print(data.values)
print(data.describe())
print(data.sort_values(by='DO'))
print(data.Temp > 26.0)
```

The first print function call outputs the names of the columns from the file, and the output is:

```python
Index(['Location', 'pH', 'DO', 'Temp'], dtype='object')
```

The second print function outputs the values from the Data Frame as a numpy array. Pandas is tightly coupled to numpy and uses numpy arrays extensively. The third print function call using pandas' `describe()` function to calculate statistics for each column of data. The output from this function is:

	Location	pH	DO	Temp
count	33.000000	33.000000	33.000000	33.000000
mean	3.606061	8.297879	9.475758	26.274545
std	1.853028	0.143891	0.991293	0.418510
min	1.000000	7.880000	7.780000	25.780000
25%	2.000000	8.240000	8.750000	26.010000
50%	4.000000	8.300000	9.520000	26.190000
75%	5.000000	8.350000	9.890000	26.420000
max	6.000000	8.650000	12.130000	27.830000

We see the average, standard deviation, and other statistical information for each column of data. The last two print functions demonstrate pandas' ability to quickly analyze and sort data. The sort_values() function reorders the rows of data based on sorting a specified column. The final print function outputs a Boolean vector that is True for every row with a temperature value over 26.0.

One of the especially useful features of pandas is that it allows us to access data using labels instead of always needing to remember column or row numbers. If the DO data set is imported into a numpy array using genfromtxt() or one of the other methods that were considered before pandas, then we need to remember that the temperature data is in the fourth column and can be accessed using data[:,3] or similar notation. If the data is stored in a pandas Data Frame, we can access a numpy array holding the temperature data using data.Temp. This is illustrated in the code below, which generates a histogram of the temperature data.

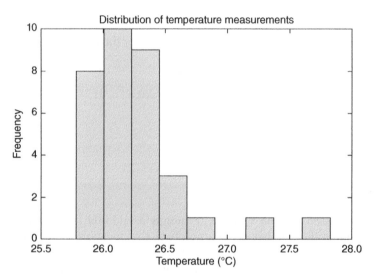

Figure 9.2 A histogram of 33 temperature measurements. The data does not appear to be normally distributed.

```
import numpy
import pandas
import matplotlib.pyplot as plt

data = pandas.read_csv("../../data/DOdata.csv")

# Plot a histogram of temperatures
plt.hist(data.Temp,9)
```

```
plt.title("Distribution of Temperature Measurements")
plt.xlabel("Temperature ($^oC$)")
plt.ylabel("Frequency")
plt.show()
```

The result is shown in Figure 9.2.

9.2.4 Parsing an Array

Once all the data from the file is imported into a numpy array, it is some-times desirable to extract out a subset of this data. For example, the data set that has been used in this chapter contains multiple pH, DO, and tempera-ture measurements at six different locations. If we want to compare the aver-age DO at location 1 to the average DO at location 2, it is helpful to parse or extract that data from the larger array. This process often requires that we first count the amount of data that is going to be extracted, then allocate space in an empty array for the data, and finally reading and copying the data into the new array.

For the data set shown in Figure 9.1, a simple function can be written that parses the larger array and returns a small array with only the location data of interest.

```
def getDOlocation(data, loc):
    countLocation = numpy.count_nonzero(data[:,0] == loc)
    DOdata = numpy.zeros(countLocation)
    countLocation = 0
    for i in range(numRows):
        if(data[i,0] == loc):
            DOdata[countLocation] = data[i,2]
            countLocation += 1
    return DOdata
```

This function is passed the full data set (a numpy array) and the location of interest. It counts the number of data points at that location, creates an empty array of appropriate size, copies the data into the new array, and then returns the data of interest. Notice the use of an equality comparator that returns a Boolean (`True` or `False`) for every row in the data set based on whether or not the location matches the variable `loc` and then the number of matching locations is counted using the `numpy.count_nonzero()` func-tion. Finally, a **for** loop is used to copy the rows matching the location into a new array.

9.3 Statistical Analysis

Before conducting any statistical analysis on a data set, it is always a good idea to plot a histogram of the data. Matplotlib (i.e., pylab) has a built-in function for generating a histogram plot from a vector containing the data of interest. The plotting function is called with: `matplotlib.pyplot.hist(data[:,3],9)` or `pylab.hist(data[:,3],9)` where the first argument is a vector with the data of interest (in this case we are asking for a histogram of the temperature data across all locations) and the second, optional argument is number of bins (i.e., the number of temperature ranges). The histogram of the temperature data is shown in Figure 9.2 and the x-axis indicates the temperature ranges for each bin and the y-axis is the frequency of data points within each range. Some of the statistical analysis that is covered in this section relies on data that is normally (or Gaussian) distributed. Examination of the histogram can provide some insight into whether or not a particular data set is normally distributed.

Numpy has built-in functions for calculating the mean and standard deviation for an array. Using the `getDOlocation()` function above to extract out the DO data for locations 1 and 2, the mean and standard deviation can be calculated using the `numpy.mean()` and `numpy.std()` function as illustrated in the code below.

```
DOdata1 = getDOlocation(data, 1)
DOdata2 = getDOlocation(data, 2)
print("Location 1: average = ", DOdata1.mean())
print(" and std. dev. = ", DOdata1.std())
print("Location 2: average = ", DOdata2.mean())
print(" and std. dev. = ", DOdata2.std())
```

For the DOdata.csv file data, the average DO for location one is 9.26 $\pm$ 0.57 mg/L and the average for location two is 9.6 $\pm$ 0.50 mg/L. An important question is whether or not these two sample means are statistically different. If additional data is collected, would the means converge to the same mean or different means? This question can be answered using a t-test, which is used to compare two means in order to determine the probability that the means are the same. For these two data sets, DOdata1 and DOdata2, a t-test comparison can be performed using `t, prob = scipy.stats.ttest_ind(DOdata1,DOdata2,equal_var = True)`. This function requires two input vectors containing the two different sets of data whose means are being compared in a t-test. A third, optional input is a Boolean (i.e., True or False) that indicates whether or not the variance is approximately equal between the two sets of data. If the two data sets are both from experimental measurements, we typically assume that their variances are equal. However, if a set of experimental data is being compared to published data, then `equal_var=False` should be used instead.

For the DO data from locations 1 and 2 in a river, the value for "t" or the t-stat is -0.95. In order to interpret this value, it needs to be compared to a critical t value from a table. We can avoid this work by simply focusing on the value of "prob" or probability that is returned. This value reflects the probability that the two means are the same. If `prob` < 0.05, then we can say with 95% confidence that the two means are different. If `prob` < 0.01 then we can be 99% confident that the means are different. These are the two thresholds commonly used in statistically comparing two data sets. For the data examined here, prob equals 0.37 so we cannot say that the two means are different and no statistical conclusion should be made. The means might be the same or they might be different, we simply do not have enough data to have confidence in either result.

In some cases, the data that we collect and want to analyze is subjective. For example, imagine if we asked 100 people to score 20 different movies from this past year on a scale of 1–10. Some people are "easy graders" and would give scores closer to 10 even when they did not really like the movie. Other people are "tough graders" and they never give a score over 8. Further, some people would use the full range of possible values, giving the worst movie a 1 and the best movie a 10 while other people would use a small, clustered range of scores (i.e., the worst movie gets a 6 and the best movie gets a 9). Whenever we have subjective scores like this with variable distributions, it is helpful to normalize the scores using a z-score, which is also known as the standard score. The z-score is defined as

$$z = \frac{x - \mu}{\sigma}, \tag{9.1}$$

where μ is the average score and σ is the standard deviation for one set of scores. Thus, each persons z-scores for the 20 movies are centered around zero (good movies have a z-score greater than zero and below average movies have a negative z-score) and the z-score reflects how many standard deviations better or worse than the average a movie ends up being rated by each individual. A z-score of 2.0 reflects a movie that is 2 standard deviations better than average (i.e., the person ranked the movie in their personal top 5% of movies).

A data set can be automatically translated into z-scores using the function: `zscore = scipy.stats.zscore(DOdata1)`. The input into the function is the original data, and the function returns the normalized scores or z-scores. The length of the **zscore** vector should be the same as the vector containing the original data, and the z-scores should average zero. The z-scores for the DO data at location 1 are as follows: -1.63, -0.65, 1.16, -0.51, 0.85, and 0.77. The greatest deviation from the average for the DO data is -1.6 standard deviations below average.

9.4 Advanced Linear Regression

Linear regression was discussed previously, but the `scipy.stats` library provides a linear regression function that returns additional information that can be helpful in interpreting the results. To explore the linear regression function in `scipy.stats`, a new data set that contains DO and temperature measurements as a function of time is used. Further, the measurements were repeated so two different time- dependent data sets are available. The data is stored in two separate csv files: DOdepletion1.csv and DOdepletion2.csv. Whenever such a situation arises, it is helpful to write a Python script to open a file and import the data for a specific trial number – trial 1 or trial 2 in this case. The Python script below is passed a trial number and it then builds the file name using the trial number and string concatenation.

```python
def loadData(trial):
    filename = 'DOdepletion' + str(trial) + '.csv'
    with open(filename, 'r') as dataFile:
        reader = csv.reader(dataFile)
        header=next(reader)
        numCols = len(header)
        maxRows = 10
        data = numpy.zeros((maxRows,numCols))
        rowCount = 0
        for row in reader:
            data[rowCount,:] = \
                numpy.array(row).astype(float)
            rowCount += 1
            if(rowCount >= maxRows):
                break
        data = numpy.resize(data, (rowCount,numCols))
        return data
```

The DO measurements from the first data set are plotted in Figure 9.3. It is clear that the DO concentration is reduced over time, but it is not clear what order process is governing the clearance. Most kinetic processes like this are first or second order. The rate of change (i.e., derivative) for a first-order process is governed by

$$\frac{dDO}{dt} = -k \cdot DO, \tag{9.2}$$

where k is the rate constant and DO the dissolved oxygen concentration in mg/L. For a first-order process, the rate of depletion is proportional to the concentration and a faster depletion rate is observed at higher concentrations. In a differential equations course or a kinetics course, the solution to this equation is derived. The solution is just presented here:

$$DO = DO_0 \cdot e^{-kt} \tag{9.3}$$

or

$$\ln(DO) = \ln(DO_0) - kt, \tag{9.4}$$

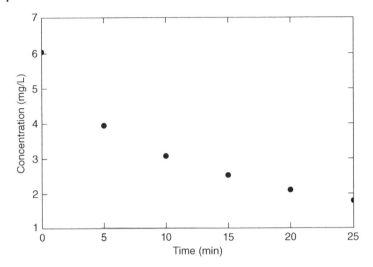

Figure 9.3 Dissolved oxygen (mg/L) measurements as a function of time for an isolated sample. The order of the depletion rate is not clear.

where DO_0 is the initial DO concentration. If a process is first order, then these equations should fit the data. On the other hand, if a process is second order, then the rate of change is governed by

$$\frac{dDO}{dt} = -k \cdot DO^2 \tag{9.5}$$

and the solution to this equation is given by

$$\frac{1}{DO} = \frac{1}{DO_0} + kt. \tag{9.6}$$

For the DO measurements plotted in Figure 9.3, we wish to determine if the process is first or second order using linear regression to fit the data to both rate laws and determining which provides the better fit.

In order to fit the data with a first-order model, we need to perform linear regression on $\ln(DO)$ versus t. Since

$$\ln(DO) = \ln(DO_0) - kt, \tag{9.7}$$

the slope from regression should be an estimate for k and the intercept should be an estimate for $\ln(DO_0)$. The `scipy.stats.linregress(x,y)` function must be passed at least two vectors as inputs: (1) the independent variable data, x, and the dependent variable data, y. For the first-order depletion model, the independent variable is time, t, and that data is in the first column of the DOdata array, and the dependent variable is $\ln(DO)$, which can be obtained by taking the natural logarithm of the second column of DOdata.

The `scipy.stats.linregress()` function can be called and passed the time and natural log of DO data using

```
slope, intercept, r_value, p_value, std_err = \
scipy.stats.linregress(dataSet1[:,0],
numpy.log(dataSet1[:,1]))
```

This function returns the slope, intercept, R, and the p-value resulting from linear regression analysis. The value of R is a measure of how close the data is to falling on a straight line. An R value close to 1.0 indicates that the data is highly linear and falls close to the regression line. The p-value indicates that probably that slope is zero, that is, the probability that the data in the second vector (the y-data) is independent of the data in the first vector (the x-data). For the first set of DO data, the value of $R^2 = 0.96$, indicating a good but not great fit between the linear regression line and the data, and the p-value is 0.0005, indicating that the DO value has an extremely low probability of being independent of time. Figure 9.4 shows the impact of adding the curve corresponding to the first-order model to the previous plot of the data. The fit is far from perfect and indicates that the first-order model may not be correct.

Fitting the data with a second-order model requires performing linear regression on $\frac{1}{DO}$ versus t. The resulting slope from linear regression should correspond to k, and the intercept should be equal to $\frac{1}{DO_0}$. The `scipy.stats.linregress()` function is called using:

```
slope, intercept, r_value, p_value, std_err = \
scipy.stats.linregress(dataSet1[:,0],1/dataSet1[:,1])
```

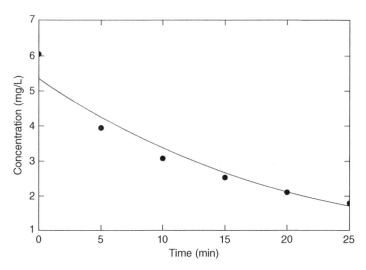

Figure 9.4 Dissolved oxygen (mg/L) measurements as a function of time and the best fit curve based on a first-order depletion model.

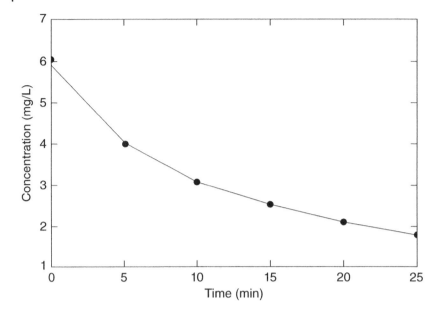

Figure 9.5 Dissolved oxygen (mg/L) measurements as a function of time and the best fit curve based on a second-order depletion model.

The value of R^2 for the second-order model is 0.999, indicating that the data falls extremely close to a straight line. This is strong evidence that the second-order model is the correct model. Once again, the p-value is very small at 3×10^{-7}. Adding the best-fit curve corresponding to the second-order model (Figure 9.5) to the original data plot shows that the model fits the data very well and the true order of the process is probably second order.

9.5   U.S. Electrical Rates Example

The goal of this final section is to review many of the tools that were introduced in this chapter through a final example. The first step is to visit www.data.gov and search for "US electric utility companies and rates." The top results for that search are links to CVS data sets compiled by the National Renewable Energy Laboratory (NREL), and you should download one of the data set. The 2011 data set is used in the example below. Analyzing electrical rates can be critical when choosing the location for an industrial facility that uses a chemical or biological process. The most extreme example might be aluminum manufacturing, which uses the Hall–Heroult process to convert aluminum oxide into pure aluminum and requires large quantities of electricity, but electrical rates are a consideration in most industrial citing decisions.

The first step in developing a Python script for exploring and analyzing this data is to import the libraries that will be used and then read in the cvs file. The example below uses the Pandas library for reading the file and storing the data.

```python
import pandas as pd
import numpy
import matplotlib.pyplot as plt
import scipy.stats

rates = pd.read_csv('iouzipcodes2011.csv')
print(rates.columns)
```

The dataset contains nine columns:

```python
['zip', 'eiaid', 'utility_name', 'state',
 'service_type', 'ownership',
 'comm_rate', 'ind_rate', 'res_rate']
```

To get a more complete picture of the data that is in the file, it is useful to next use the function call: `print(rates.describe())`. Some of the information output by this function includes the following:

- There are 37,791 rows of data in the file, indicating that there are 37,791 unique zip codes.
- The average commercial rate (`comm_rate`) is $0.084/kWh.
- The average industrial rate (`ind_rate`) is $0.063/kWh.
- The average residential rate (`res_rate`) is $0.103/kWh.
- The standard deviation for all rate levels is approximately $0.04/kWh.
- The maximum residential rate is $0.85/kWh.

The very high maximum rates for each of the rate categories are interesting. Generating a quick scatter plot of the residential rate versus zip code provides some insight.

```python
plt.plot(rates.zip,rates.res_rate,'o')
plt.xlabel('zip code')
plt.ylabel('residential rate')
print('Zip Code with highest rate: ',
      rates.zip[numpy.argmax(rates.res_rate)])
```

The plot that is generated is shown in Figure 9.6. Most of the residential electric rates for the United States are less than $0.20/kWh, but there are a few extreme outliers. The last line of the Python code from the section shown above uses the numpy.argmax() function to get the row number with the largest value for the residential rate and then it prints out the zip code associated with that row number. The zip code is 99634, and a quick search shows that this is the zip code for Napakiak, Alaska.

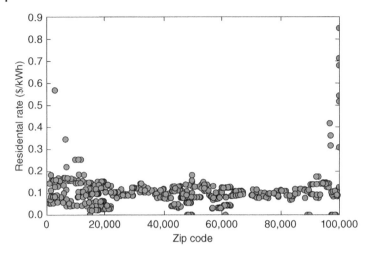

Figure 9.6 Scatter plot showing the residential electric rate ($/kWh) for every U.S. zip code.

This is a very large data set that covers the entire United States, but we might want to focus in on a specific state. The next section of the Python script counts up the number of rows in the data set associated with the state of Texas and then builds numpy arrays that hold the residential and commercial electric rates for Texas only.

```
TXcount = numpy.count_nonzero(rates.state == 'TX')
TXcomm = numpy.zeros(TXcount)
TXres = numpy.zeros(TXcount)
count = 0
for i in numpy.nditer(numpy.nonzero(rates.state == 'TX')):
    TXcomm[count] = rates.comm_rate.data[i]
    TXres[count] = rates.res_rate.data[i]
    count += 1
```

A numpy function that has not been used before appears in the code above. The numpy.nditer() function converts a numpy array into a Python object that can be iterated over, allowing a **for** loop to loop over all the rows for the state of Texas. Select data in those rows for Texas is then copied into smaller, Texas-specific arrays.

Finally, the final section of the Python script (below) generates a histogram of the various residential electric rates in Texas:

One question concerning electrical rates might be whether there is a correlation between the residential rate and the commercial rate. Figure 9.7 shows the commercial rate (x-axis) versus residential rate (y-axis) for the zip codes in Texas. Even though there are a large number of zip codes for Texas (456), there are a much smaller number of unique residential and commercial rates. It is also clear from Figure 9.7 that there is a strong, but not perfect, correlation

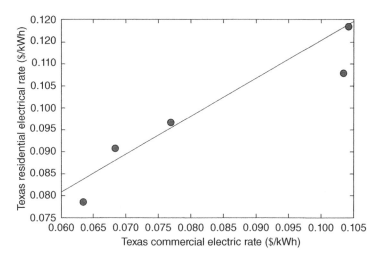

Figure 9.7 Scatter plot showing the commercial electric rate ($/kWh) versus the residential electric rate ($/kWh) for every Texas zip code.

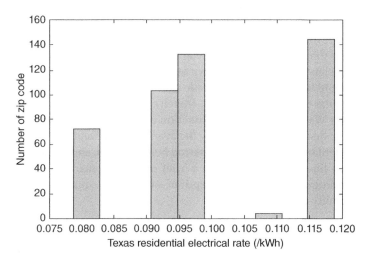

Figure 9.8 Histogram showing the frequency of zip codes with various residential electric rates ($/kWh).

between the two rates. We can go one step further by calculating the R^2 value for a linear regression function through this data. Using

```
slope, intercept, r_value, p_value, std_err = \
    scipy.stats.linregress(TXcomm,TXres)
```

it can be found that the residential rate is equal to 0.87 times the commercial rate plus $0.03) and the $R^2 = 0.98$, indicating a strong correlation.

```
plt.hist(TXres,10)
plt.xlabel('Texas Res. Elec. rate')
plt.ylabel('number of zip codes')
```

and the resulting figure is shown in Figure 9.8. Interestingly, the highest rate in Texas ($0.12/kWh) is also the most frequent rate based on zip code.

Problems

9.1 You have been hired to statistically analyze the snow pack for one basin (or area) in Montana over the past decade. To begin, go to the Natural Resources Conservation Service's snow data (SNOTEL) website containing the Monthly Historic Snow (SNOTEL) Data for various basins in Montana: http://www.wcc.nrcs.usda.gov/nwcc/rgrpt? report=snowcourse&state=MT

Your employer has made the unusual request of asking you to analyze the annual Snow Data from a random site of your choosing. The website will allow you to download a text file containing the snow pack data for various months going back different lengths in time depending on the year. The text file follows the comma separated value (*.csv) format with the exception of the header information. I recommend downloading the data, changing the file extension to csv and opening the file in Excel so that you can delete any unwanted information. If you do open the file in Excel, you should be sure to resave the file as a *.csv file.

Select a particular time of the year (e.g., March), and plot either the Snow Depth or Snow Water Equivalent as a function of year for at least the last 10 years. Add a line to the plot showing the average Snow Depth or Snow Water Equivalent over the entire time period being plotted, and add standard deviation lines if possible. Write a brief report summarizing your findings regarding the average snow depth or snow water equivalent and the variiings ability from year to year.

9.2 You have been hired by a major clothing manufacturing to provide some analysis and figures (plots) on historical clothing sales trends and clothing advertising spending. Begin by downloading publically available clothing expenditure data from http://www.stat.ufl.edu/winner/data/clotthes_expend.csv and the description of the data (optional): http://www.stat.ufl.edu/winner/data/clothes_expend.txt. The csv data file contains a column of data on the real US GDP in billions of dollars that include quotation marks around the number. It is recommended that this column be deleted by opening the file in Excel before importing the file into a Python array. After importing the data into Python, you are asked to generate a plot of clothing sales (in billions of dollars, second column of data) for every year that data is available. You are also asked to generate a second plot showing expenditures on advertising as a percent of GDP (last column) along with the average percentage and standard deviation over the full time period. Finally, returning to the original clothing sales (in billions of dollars, second column of data) per year data, use linear regression to determine the slope of the curve over the *most recent 10 years of data*. The slope will represent the average increase in clothing sales per year for the last 10 years. Write a brief memo to the clothing manufacturer that summarizes your findings on the changes in sales, sales trends, and average advertising spending as a percentage of GDP.

9.3 In 2015, the New England Patriots were accused of illegally deflating the footballs that they used on offense during the AFC Championship game. The *Wells Report* was released in May 2015, and contains the findings of a special commission hired by the NFL to investigate the accusations. The primary finding of the commission was, "it is more probable than not that New England Patriots personnel participated in violations of the Playing Rules and were involved in a deliberate effort to circumvent the rules." In other words, it is statistically probable that the Patriots deflated their footballs below the 12.5 psig minimum allowed by the Playing Rules.

The primary pressure data used to make this determination is from the pressure measurements of two different referees using two different pressure gauges at halftime. It is assumed by the report that the Patriots footballs were all inflated to 12.5 psig at the start of the game and before tampering. The Colts footballs were all inflated to 13.0 psig at the start of the game. (It is unclear which pressure gauge was used at the start of the game.) The halftime pressure data that was collected is summarized in the table below (all pressure data is in psig).

Patriots Ball	Blakeman	Prioleau
1	11.50	11.80
2	10.85	11.20
3	11.15	11.50
4	10.70	11.00
5	11.10	11.45
6	11.60	11.95
7	11.85	12.30
8	11.10	11.55
9	10.95	11.35
10	10.50	10.90
11	10.90	11.35

Colts Ball	Blakeman	Prioleau
1	12.70	12.35
2	12.75	12.30
3	12.50	12.95
4	12.55	12.15

You have been hired by the New England Patriots to repeat some of the statistical analysis presented in the Wells Report and then submit your own expert opinion. Specifically, calculate average pressure *decrease* (and standard deviation) for the Patriot's and Colt's balls based on the measurements of each pressure gauge. A *t*-test is used to compare two means (or two average) and determine the probability that they are the same. Use a *t*-test (`scipy.stats.ttest_ind()` is recommended) to determine the probability that the average pressure drop of the Patriot's footballs is different from the pressure drop of the Colt's footballs for each set of measurements. (Extra Credit) The Patriots have offered you a bonus payment if you are able to generate a bar chart showing the average pressure drop and standard deviation for the two different sets of measurements for their footballs.

Finally, the Patriots have asked you to summarize your own expert opinion based on the data and facts presented here. Do you think there is strong evidence of deflation in violation of the Playing Rules. Do not consider facts beyond the pressure data (e.g., text messages included in the report).

9.4 You have been hired by a local lawyer that is defending an individual accused of driving under the influence of alcohol. The defendants argument is that the breathalyzer test used to measure the individuals blood alcohol concentration (BAC) is not reliable or consistent. The lawyer is relying on data from a previously published study by Gullberg [2]. In this study, 10 breath alcohol samples were taken from different subjects approximately 20 s apart. The defense attorney's argument is that a reliable test should give the same measurement for all 10 samples, but this study showed some variability.

Your specific task as a consultant is to generate figures that the lawyer can use during the trial. The figures should show the BAC measurement for each of the 10 samples from a single subject as well as the average and standard deviation lines for the samples from the individual. The lawyer has requested figures for at least two different subjects. A csv file containing the data for two different subjects is available in the paper referenced above or http://www.stat.ufl.edu/ winner/data/breath_reg.dat.

To help you generate exactly the plot desired, the lawyer drew a sketch on a napkin shown below. In addition to the BAC figures, the lawyer also needs your report to include the averages and standard deviations for the data from the two subjects (Figure 9.9).

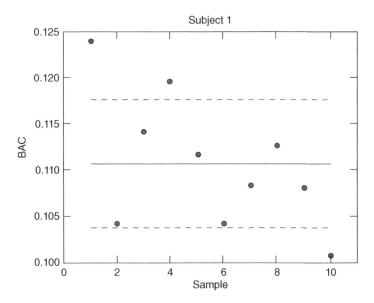

Figure 9.9 Rough sketch of desired BAC figure.

References

1 McKinney, W. (2012) *Python for Data Analysis*, O'Reilly Media, Inc., Sebastopol, CA.

2 Gullberg, R.G. (1995) Repeatability of replicate breath alcohol measurements collected in short time intervals. *Sci. Justice*, **35**, 5–9.

10

Numerical Differentiation and Integration

10.1 Introduction

Most students learn how to calculate the derivative or integral of a given function in a first-year calculus course. Computational methods for taking derivatives and integrals of functions were also described in Chapter 5 on symbolic computations. For some complex functions, however, it is simply not possible to calculate the integral or even the derivative, and it may be necessary to obtain a numerical approximation of the derivative or integral. Sometimes, it is just easier to use a numerical approximation than to determine the exact derivative or integral. In this chapter, standard approaches for numerically approximating derivatives and integrals are described. The numerical approximation of a derivative is usually not particularly useful until it is used to approximately solve differential equations, as described in the later chapters. The numerical approximation of a definite integral, however, is immediately useful for calculating quantities such as enthalpy changes and reactor volumes. This chapter is divided into two parts. The first part briefly examines numerical approximations for the first and second derivatives of a given function at a point. The second part describes a few different methods of numerically approximating definite integrals.

10.2 Numerical Differentiation

The numerical approximation of a derivative has already been briefly explored when Newton's and Broyden's methods were discussed earlier. In this section, common choices for numerically approximating a derivative are examined along with an analysis of the error associated with the approximation. These approximations can be derived from the Taylor polynomial, but here, the Taylor polynomial will be written in a slightly different form from that shown in Chapter 8 on nonlinear equations. The Taylor polynomial provides an approximation for a function about the point x_i. The value of the function a

Chemical and Biomedical Engineering Calculations Using Python®, First Edition. Jeffrey J. Heys.
© 2017 John Wiley & Sons, Inc. Published 2017 by John Wiley & Sons, Inc.
Companion Website: www.wiley.com/go/heys/engineeringcalculations_python

distance h away from x_i is

$$f(x_i + h) = f(x_i) + h\frac{df(x_i)}{dx} + \frac{h^2}{2}\frac{d^2f(x_i)}{dx^2} + \cdots \tag{10.1}$$

or

$$f(x_i + h) = f(x_i) + h\frac{df(x_i)}{dx} + O(h^2), \tag{10.2}$$

where $O(h^2)$ represents a term with a size on the order of h^2. Equation 10.2 can be seen to be equivalent to the previous Taylor polynomial equation 8.3 by replacing h with $x - x_i$.

10.2.1 First Derivative Approximation

Our goal in this section is to approximate $\frac{df(x_i)}{dx}$, and we can rearrange the Taylor polynomial (equation 10.2) to give

$$\frac{df(x_i)}{dx} = \frac{f(x_i + h) - f(x_i)}{h} + O(h) \tag{10.3}$$

or

$$\frac{df(x_i)}{dx} \approx \frac{f(x_i + h) - f(x_i)}{h}. \tag{10.4}$$

This approximation is typically referred to as the forward approximation of the first derivative, and the error associated with the approximation is of order h. As we will see in the next few chapters, numerical approximation of derivatives is often performed on a sequence of points that are separated by a fixed distance h. Hence, it is common to simplify this notation slightly by replacing $f(x_i)$ by f_i and $f(x_i + h)$ with f_{i+1}, see Figure 10.1, so that it clearly refers to the next point in a sequence of points. Using this notation, the forward difference approximation of the derivative becomes

$$\frac{df(x_i)}{dx} \approx \frac{f_{i+1} - f_i}{h}. \tag{10.5}$$

The use of this approximation is shown in Figure 10.1(b). Clearly, the approximation of the slope becomes more accurate as the distance between the two points shrinks toward zero.

The Taylor polynomial at $f(x_i - h)$ is

$$f(x_i - h) = f(x_i) - h\frac{df(x_i)}{dx} + O(h^2) \tag{10.6}$$

and using the same derivation as above, leads to the backward difference approximation of the derivative:

$$\frac{df(x_i)}{dx} \approx \frac{f_i - f_{i-1}}{h}, \tag{10.7}$$

which has the same $O(h)$ accuracy as the forward difference approximation. The use of the backward difference approximation is shown in Figure 10.1(a).

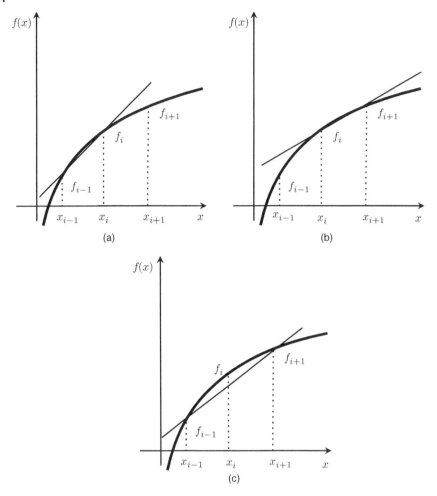

Figure 10.1 Three different finite difference approximations for the derivative at x_i: (a) backward difference, (b) forward difference, and (c) centered difference.

The final method for approximating the derivative of $f(x)$ at x_i is by subtracting the Taylor polynomial for $f(x_i - h)$ (equation 10.6) from the polynomial for $f(x_i + h)$ (equation 10.2). Notice that the $f(x_i)$ terms are eliminated and, more importantly, it can be shown that the $O(h^2)$ terms are eliminated leaving a $O(h^3)$ term. The resulting centered difference approximation of the first derivative is

$$\frac{df(x_i)}{dx} \approx \frac{f_{i+1} - f_{i-1}}{2h},\tag{10.8}$$

which is $O(h^2)$ accurate. As a result, as $h \to 0$, the centered difference approximation is much more accurate than either the forward or backward difference approximation. The centered difference approximation is illustrated in Figure 10.1(c).

To illustrate both the implementation and accuracy of numerical approximations for the first derivative, the Python code below was created. The code has the function $f(x) = x \cdot \sin(x)$ declared near the beginning, and the code has a second function that calculates the exact derivative. The derivative at $x = 1.0$ is approximated using the three different approximations described above (equations 10.5, 10.7, 10.8), and the accuracy of each approximation is calculated.

```python
import math

def fun(x):
    return x*math.sin(x)

def exactDeriv(x):
    return math.sin(x)+x*math.cos(x)

x0 = 1.0
exact = exactDeriv(x0)

for i in range(5):
    h = 10**(-i-1)
    df_forw = (fun(x0+h)-fun(x0))/h
    df_back = (fun(x0)-fun(x0-h))/h
    df_center = (fun(x0+h)-fun(x0-h))/(2*h)
    print("forward error: %.1e" %
            math.fabs(df_forw-exact))
    print("backward error: %.1e" %
            math.fabs(df_back-exact))
    print("centered error: %.1e" %
            math.fabs(df_center-exact))
```

Running the Python script for numerical differentiation gives the results shown in Table 10.1. The forward and backward finite difference approximations give very similar levels of error, and both options give an error that is $O(h)$. The centered difference approximation, on the other hand, has a similar level of error for large values of h, but it converges much faster to the exact solution and has an error that is $O(h^2)$. Interestingly, even though it is not clear at this point why anyone would ever consider using a forward or backward difference approximation, we will later see situations where we need to accept the poor accuracy of the lower order approximations because they demonstrate better numerical stability. Therefore, do not forget about the forward and backward difference approximations because they will have a use later.

Table 10.1 Accuracy of different numerical approximations of the first derivative of $f(x) = x \cdot \sin(x)$

h	Forward difference error	Backward difference error	Centered difference error
0.1	6.8×10^{-3}	1.7×10^{-2}	5.1×10^{-3}
0.01	1.1×10^{-3}	1.2×10^{-3}	5.1×10^{-5}
0.001	1.2×10^{-4}	1.2×10^{-4}	5.1×10^{-7}
0.0001	1.2×10^{-5}	1.2×10^{-5}	5.1×10^{-9}
1.0×10^{-5}	1.2×10^{-6}	1.2×10^{-6}	5.5×10^{-11}

10.2.2 Second Derivative Approximation

The second derivative of a function can also be numerically approximated, and, once again, we turn to the Taylor polynomial to derive an equation for the approximation. Recalling that

$$f(x_i + h) = f(x_i) + h\frac{df(x_i)}{dx} + h^2\frac{d^2f(x_i)}{dx^2} + O(h^3) \tag{10.9}$$

and

$$f(x_i - h) = f(x_i) - h\frac{df(x_i)}{dx} + h^2\frac{d^2f(x_i)}{dx^2} - O(h^3), \tag{10.10}$$

we can add these two equations (10.9 and 10.10) together (note that the first derivative terms cancel and the $O(h^3)$ terms cancel) giving

$$f(x_i + h) + f(x_i - h) = 2f(x_i) + h^2\frac{d^2f(x_i)}{dx^2} + O(h^4). \tag{10.11}$$

This equation can be rearranged to solve for the second derivative and the notation simplified to yield

$$\frac{d^2f(x_i)}{dx^2} \approx \frac{f_{i+1} - 2f_i + f_{i-1}}{h^2}, \tag{10.12}$$

which is $O(h^2)$ accurate. This approximation is also called the centered difference approximation for the second derivative. There are other approximations that have been derived for the second derivative, but this one approximation is used in the vast major of engineering algorithms in the author's experience.

The accuracy of the centered difference approximation (equation 10.12) of the second derivative is examined using the Python script below.

```
import math

def fun(x):
    return x*math.sin(x)
```

```
def exactSecDeriv(x):
    return 2.0*math.cos(x)-x*math.sin(x)

x0 = 1.0
exact = exactSecDeriv(x0)

for i in range(5):
    h = 10**(-i-1)
    ddf_center = (fun(x0+h)-2.0*fun(x0)+fun(x0-h))
    ddf_center = ddf_center/(h**2)
    print("h = ", h, ": error is %.4e" %
        math.fabs(ddf_center-exact))
```

The output from this script is:

```
h = 0.1 : error is 1.0991e-03
h = 0.01 : error is 1.0998e-05
h = 0.001 : error is 1.1005e-07
h = 0.0001 : error is 2.9147e-09
h = 1e-05 : error is 1.7512e-06
```

For this example, the accuracy of the approximation is $O(h^2)$ until h is reduced to $h = 1 \times 10^{-5}$, at which point computer roundoff error (or floating point truction error) limits further reduction in approximation error for the second derivative.

10.2.3 Scipy Derivative Approximation

The scipy library includes a function, `scipy.misc.derivative()`, for numerically approximating derivatives. The inputs into the function are:

func a required input function whose derivative will be approximated
x0 the required point at which the derivative is approximated
dx *optional* spacing between the differencing points
n *optional* order of the derivative, default is 1
order *optional* number of points to use, must be odd

The use of the `scipy.misc.derivative()` function along with the centered difference formulas derivative above (equations 10.8 and 10.12) is demonstrated in the Python script below, which approximates the derivative of $f(x) = 2x \cdot e^x$ at $x = 0.5$.

```
import math
import scipy.misc

def fun(x):
    return 2.0*x*math.exp(x)
```

```
def exactDeriv(x):
    return 2.0*(1+x)*math.exp(x)

def exactSecDeriv(x):
    return 2.0*(2.0+x)*math.exp(x)

x0 = 0.5
exact = exactDeriv(x0)
exactSec = exactSecDeriv(x0)

print("Scipy deriv error is %.4e" %
      math.fabs(scipy.misc.derivative(fun,x0)
      - exact))
print("Scipy second deriv error is %.4e" %
      math.fabs(scipy.misc.derivative(fun,x0,n=2)
      - exactSec))

for i in range(3):
    h = 10**(-i-2)
    # forward difference
    df = (fun(x0+h)-fun(x0-h))/(2*h)
    ddf = (fun(x0+h) - 2.0*fun(x0) + fun(x0-h))
    ddf = ddf/(h**2)
    print("h =",h,": deriv. error is %.4e"
          % math.fabs(df-exact))
    print("             second deriv. error is %.4e"
          % math.fabs(ddf-exactSec))
```

The output from this script is:

```
Scipy deriv error is 2.0796e+00
Scipy second deriv error is 1.2975e+00
h = 0.01 : deriv. error is 1.9235e-04
            second deriv. error is 1.2365e-04
h = 0.001 : deriv. error is 1.9235e-06
            second deriv. error is 1.2364e-06
h = 0.0001 : deriv. error is 1.9234e-08
            second deriv. error is 2.1802e-08
```

The very large error for the `scipy.misc.derivative()` function approximation of the first and second derivatives indicates that the function is probably being used incorrectly. Reflecting back on the inputs to the function, we can observe that one of the optional inputs is dx, which is the spacing between the differencing points, that is, dx is the same as h in the difference approximation derived here. Reviewing the available documentation on the `scipy.misc.derivative()` function reveals that the default value for dx is 1.0, which is very large for this particular example problem. Modify the function calls to the `scipy.misc.derivative()` function to be:

```
print("Scipy deriv error is %.4e" %
      math.fabs(scipy.misc.derivative(fun,x0,dx=0.0001)
      - exact))
print("Scipy second deriv error is %.4e" %
      math.fabs(scipy.misc.derivative(fun,x0,
      dx=0.0001,n=2) - exactSec))
```

results in the error being:

```
Scipy deriv error is 1.9234e-08
Scipy second deriv error is 2.1802e-08
```

and this is identical to the centered difference approximations presented here.

10.3 Numerical Integration

The calculation of the definite integral of a function in one dimension is identical to calculating the area between the function and the x-axis. If we were not at all concerned with accuracy, we could approximate the function with a straight line between the bounds on the definite integral, (a, b). The resulting polygon would be a trapezoid and we could easily approximate the area with

$$\int_a^b f(x)dx \approx \frac{h}{2}[f(a) + f(b)].$$

The error associated with this approximation depends on the nonlinearity of the function (i.e., roughly, the second derivative) but for most problems, the use of a single trapezoid is not sufficiently accurate and much better accuracy is possible at a modest computational cost.

If the region under the function is subdivided into n intervals of width $h = (b - a)/n$, then arbitrarily high accuracy is possible by increasing n. Approaches that use multiple polygons to approximate the area under the curve are called composite method. The simplest composite method is to divide the region under the function into rectangles. Each rectangle typically has the width, h, and the height of each rectangle is determined by evaluating the function at the midpoint of the rectangle's width (i.e., the midpoint of the first rectangle is $a + h/2$). This approach is called the composite midpoint rule because the midpoint is used to determine the area of each rectangle. The midpoint rule process is illustrated in Figure 10.2. The equation describing this approximate integral is

$$\int_a^b f(x)dx \approx h \sum_{j=0}^n f(x_j), \tag{10.13}$$

where x_j is at the midpoint of each subdomain. The error associated with this approximation is $O(h^2)$, which means that using twice as many subdomains (i.e., reducing h in half) results in a factor of 4 reduction in the error.

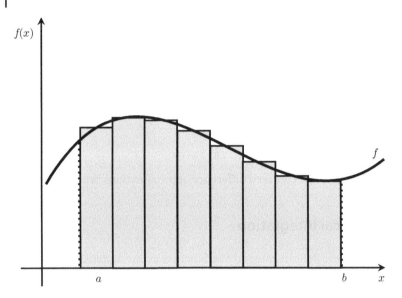

Figure 10.2 The fundamental idea behind the numerical approximation of a definite integral (from *a* to *b*) is to estimate the area under the curve using simple subdomains with areas that are easy to calculate. The composite midpoint rule is illustrated here.

A test problem using the midpoint rule (equation 10.13) to approximate $\int_0^2 x \cdot \sin(x)dx$ is solved by the Python script below.

```python
import math
import numpy

def fun(x):
    return x*numpy.sin(x)

def exactIntegral(a, b):
    integral = -b*math.cos(b)+math.sin(b)
    integral += a*math.cos(a) - math.sin(a)
    return integral

a = 0.0
b = 2.0

exact = exactIntegral(a, b)

# Midpoint Rule
n = 100
h = (b-a)/n
x = numpy.linspace(a+h/2,b-h/2,num=n)
area = 0
```

```
for i in range(n):
    area = area + h*(fun(x[i]))
print("Midpoint rule error: %.4e"
        % math.fabs(exact-area))
```

The one line in this Python code that is somewhat more complex than the others is the construction of the vector, **x**, that holds the midpoints. The key in constructing this vector is to start at the midpoint of the first interval $(a + h/2)$ and linearly space the points to the midpoint of the last interval $(b - h/2)$. Using 100 subinterval (or 100 rectangles) the midpoint rule results in an error of approximately 1.3×10^{-6}, which is sufficiently accurate for most engineering problems.

10.3.1 Trapezoid Rule

Instead of approximating the area under the function with a sequence of rectangles, a sequence of trapezoids could be used instead. The process is illustrated in Figure 10.3. The equation describing this process is

$$\int_a^b f(x)dx \approx \frac{h}{2}\left[\sum_{j=0}^{(n-1)} f(x_j) + f(x_{j+1})\right]. \tag{10.14}$$

Interestingly, the accuracy of the composite trapezoid rule is the same as the composite midpoint rule, $O(h^2)$, but the trapezoid rule has the same overall

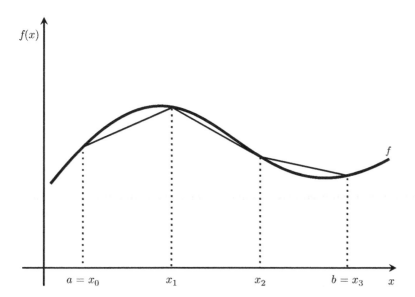

Figure 10.3 The area under a function (i.e., the definite integral of a function) can be estimated by subdividing the area into a sequence of trapezoids.

computational cost and is sometimes more accurate by a factor of 2. Whenever we read that the order of accuracy is h^2 (i.e., $O(h^2)$), we can think of this as saying that the error is equal to $k \cdot h^2$ where k is a constant, or we can think of this as saying that the error is proportional to h^2. If the error of the composite midpoint rule is $k \cdot h^2$, then the error associated with the composite trapezoid rule for that same problems is $\frac{k}{2}h^2$. Because the composite trapezoid rule provides twice the accuracy for the same computational cost, it is generally preferred over the composite midpoint rule.

The composite trapezoid rule is straightforward to implement in Python, and an (inefficient) implementation is given in the script below. This implementation is inefficient because the function being integrated is evaluated twice at the same location, x. This is obviously a waste, but the algorithm is so fast that this little inefficiency does not really matter.

```
import math
import numpy

def fun(x):
    return x*numpy.sin(x)

def exactIntegral(a, b):
    ans = -b*math.cos(b)+math.sin(b)
    ans = ans+a*math.cos(a)-math.sin(a)
    return ans

a = 0.0
b = 2.0

# Trapazoid Rule
n = 100
h = (b-a)/n
x = numpy.linspace(a,b,num=n+1)
area = 0
for i in range(n):
    area = area + h*(fun(x[i]) + fun(x[i+1]))/2.0
print(area)
```

For the test problem used here, using $n = 10$ results in an error of 3.3×10^{-3} and using $n = 100$ results in an error of 2.6×10^{-6}. This error reduction is even larger than expected ($O(h^2)$ is expected) due to the smoothness of the function being integrated.

10.3.2 Numerical Integration Using Scipy

The numerical integration techniques examined thus far are based on evaluating the function being integrated at evenly spaced points, that is, the midpoint of a subdomain or the endpoints of a subdomain. Higher accuracy can be achieved by evaluating the function being integrated at strategically placed points that minimize the error associate with the approximation. These "optimal" points are not at the ends of the subdomain, and they are not evenly spaced throughout the subdomain. Fortunately, mathematicians have previously determined the locations of these optimal points, they are called *Gauss points*, and the numerical integration method is called *Gaussian quadrature*.

An algorithm that implements the Gaussian quadrature approach for numerically approximating integrals is included in the `scipy.integrate` library. The use of this algorithm is illustrated in the Python script below.

```python
import math
import numpy
import scipy.integrate.quadrature as quad

def fun(x):
    return x*numpy.sin(x)

def exactIntegral(a, b):
    ans = -b*math.cos(b)+math.sin(b)
    ans = ans + a*math.cos(a)-math.sin(a)
    return ans

a = 0.0
b = 2.0

exact = exactIntegral(a, b)
estimate = quad(fun,a,b)
print("error: ", math.fabs(exact-estimate[0]))
```

The Gaussian quadrature function is passed the name of a Python function contain the equation we are approximately integrating, and it is passed the bounds on the definite integral. The use of Gaussian quadrature provides an extremely accurate approximation of the integral at a relatively modest computational cost in many cases. For the example equation used in the Python script above, $f(x) = x \cdot \sin(x)$, the Gaussian quadrature algorithm in Scipy returns the approximate integral that has an error of only 9×10^{-12}. This is basically as close as possible on a computer to the exact integral.

10.3.3 Error Function

The example problem that was examined in the previous codes, $\int_0^2 x \cdot \sin(x)$, has an exact solution that can be found analytically. One of the major uses of numerical integration, however, is for integrating functions that do not have an exact, analytical solution, that is, it is impossible to find an antiderivative that is an elementary function. One example is

$$\int e^{-x^2} \, dx. \tag{10.15}$$

This integrand arises frequently when solving heat conduction or diffusion problems on semi-infinite domains. For example, modeling the diffusion of a drug from a skin patch into the tissue below may result in this integrand. Another example is the modeling of soil temperatures near the surface, which oscillates due to solar heating but the oscillations decay with depth (i.e., the temperature does not change a few feet below the surface).

This integrand arises so frequently, that a "special function" (i.e., nonelementary function) has been defined. The error function is defined as

$$erf(y) = \frac{2}{\sqrt{\pi}} \int_0^y e^{-x^2} \, dx. \tag{10.16}$$

The value of this function can be obtained using tables or many software libraries having functions that provide highly accurate approximations. In Python, the `scipy.special.erf()` function is available in the `scipy` library. The Python script below approximates the integral to

$$\int_0^\pi e^{-x^2} \, dx$$

using the Midpoint rule (equation 10.13), the trapezoid rule (equation 10.14), and Gaussian quadrature. These approximations are then compared to the hopefully highly accurate approximation available using `scipy.special.erf()`.

```python
import math
import numpy
import scipy.integrate.quadrature as quad
import scipy.special as ss

def fun(x):
    return numpy.exp(-x*x)

def exactIntegral(a, b):
    integral = (ss.erf(b) - ss.erf(a))
    return math.sqrt(math.pi)*integral/2.0

a = 0.0
b = math.pi
exact = exactIntegral(a, b)
n = 100
h = (b-a)/n
```

```
# Gaussian Quadrature
estimate = quad(fun,a,b)
print("Gaussian Quadrature error: %.4e"
      % math.fabs(exact-estimate[0]))

# Midpoint Rule
x = numpy.linspace(a+h/2,b-h/2,num=n)
area = 0
for i in range(n):
    area = area + h*(fun(x[i]))
print("Midpoint rule error: %.4e"
       % math.fabs(exact-area))

# Trapazoid Rule
x = numpy.linspace(a,b,num=n+1)
area = 0
for i in range(n):
    area = area + h*(fun(x[i])+fun(x[i+1]))/2.0
print("Trapazoid rule error: %.4e"
       % math.fabs(exact-area))
```

The "exact" integral is assumed to be the highly accurate approximation from `scipy.special.erf()`, and the difference between this "exact" integral and Gaussian quadrature using `scipy.integrate.quadrature` is 5.5×10^{-11}. Using numerical integration with the midpoint rule and the trapezoid rule and as few as 10 intervals also provides a relatively accurate solution as summarized in Table 10.2.

Table 10.2 Accuracy of the midpoint and trapezoid rule for integrating $\int_0^\pi e^{-x^2} \, dx$ with different numbers, *n* of subintervals

n	Midpoint error	Trapezoid error
10	1.2×10^{-6}	3.1×10^{-6}
100	1.3×10^{-8}	2.7×10^{-8}
1000	1.2×10^{-10}	2.7×10^{-10}

Problems

10.1 The `scipy.misc.derivative()` function has an optional input argument `order` that specifies the number of points to use in the difference approximation and must be an odd number. The default value for this function is 3, and the centered difference approximations derived at the start of this chapter were also based on three points. Evaluate the impact of setting `order` to 5 and 7 while changing dx to 0.01, 0.001, and 0.0001 for the function used previously, $f(x) = 2x \cdot e^x$. What do you observe? Describe the potential advantages and disadvantages associating with using a higher order approximation.

10.2 You have been hired by a mathematics software company that has previously been using the composite trapezoid rule to numerically approximate the integrals of various mathematical functions of the form:

$$\int_a^b f(x)\,dx.$$

For the composite trapezoid rule, the integration interval, $[a, b]$, is divided into smaller pieces and then the integral of each piece is approximated using

$$\int_{x_i}^{x_{i+1}} f(x)\,dx \approx \frac{h}{2}[f(x_i) + f(x_{i+1})].$$

The software company has hired you to implement Simpson's rule, which is similar to the trapezoid rule except that instead of approximating the integral of each piece with a trapezoid, the integral of each piece is approximated using a quadratic polynomial. Thus, the integral of each piece of the interval is approximated using

$$\int_{x_i}^{x_{i+1}} f(x)\,dx \approx \frac{h}{3}[f(x_i) + 4 \cdot f(x_{i+1}) + f(x_{x+2})].$$

This approximation is illustrated in Figure 10.4. Simpson's rule should be implemented and tested by integrating $\int_0^2 x \cdot \sin(x)\,dx$ for a specified number of points, n. Note that each interval consists of 3 points, which means that there are only $\frac{n-1}{2}$ intervals (and, n must be odd so that $n = 5$ points corresponds to 2 intervals) and a loop through the intervals should have a step size of 2. Evaluate the accuracy of Simpson's rule for different values of $h = \frac{b-a}{n-1}$ the distance between the points. Is Simpson's rule more accurate than the trapezoid rule for a given h?

10.3 The calculation of the amount of energy required to change the temperature of a material is performed frequently in process engineering. For example, if we have a cubic meter of nitrogen or a kilogram of carbon,

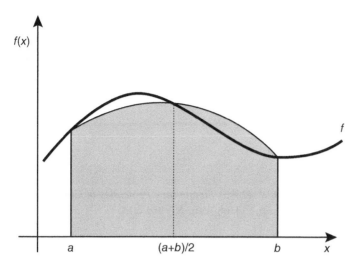

Figure 10.4 Illustration of Simpson's rule for integration of $f(x)$ from a to b.

how much energy is required to raise the temperature by 1 or 100 °C? The calculation frequently requires integrating a polynomial between the starting and ending temperatures. A refinery has hired you to create a Python program that both automates these calculations and compares two different numerical methods for the approximation of integrals. Assuming a constant pressure process, the enthalpy change associated with a temperature change from T_1 to T_2 is

$$\Delta \hat{H} = \int_{T_1}^{T_2} C_p(T) dT$$

where $C_p(T)$ is the heat capacity as a function of temperature at constant pressure [1]. The heat capacity of a material as a function of temperature is frequently given through a polynomial relationship. The refinery is particularly interested in two materials: (1) nitrogen gas and (2) carbon solid. The heat capacity of nitrogen gas in kJ/(mol · °C) is given by

$$C_p(T) = 0.0290 + 0.2199 \times 10^{-5} T + 0.5723 \times 10^{-8} T^2$$
$$- 2.871 \times 10^{-12} T^3,$$

where T should be in °C. The heat capacity of carbon solid in kJ/(mol · K) is given by

$$C_p(T) = 0.1118 + 1.095 \times 10^{-5} T + 489.1/T^2,$$

where T should be in K.

Your contract with the refinery asks you to calculate the enthalpy change for each material (nitrogen gas and carbon solid) starting at a temperature of 20°C and raising the temperature to 100°C. Report the enthalpy change in kJ/mol. You should numerically approximate the integral using the Gaussian quadrature algorithm included in Scipy (`scipy.integrate.quadrature()`). In addition, use the Trapazoid rule with 10 intervals to approximate the enthalpy change of nitrogen gas and report the difference between the trapazoid rule and Gaussian quadrature.

Reference

1 Felder, R.M. and Rousseau, R.W. (2005) *Elementary Principles of Chemical Processes*, John Wiley & Sons, Inc., Hoboken, NJ, 3rd edn.

11

Initial Value Problems

11.1 Introduction

Within the various fields of science and engineering, we are often interested in how a system that is at a known state reacts to changes in a parameter that influences that system. For example, what happens to the current in a circuit if there is a change in the resistance across the device in the circuit? What happens to the temperature of the coolant leaving a radiator if the air temperature changes? How fast is the change? What is the magnitude of the change? Many systems can be described by a differential equation that contains derivatives with respect to time. Typically, the *initial conditions* of the system are known and we are interested in modeling the long-term behavior of the system. Problems in this important category are called *initial value problems* and they are the focus of this chapter.

11.2 Biochemical Reactors

In biological systems, enzymes catalyze most of the reactions where one compound or substrate is converted into a product. For example, wine contains ethanol. In some cases, an enzyme from a microorganism can be present that converts the ethanol into acetic acid. If this happens, the wine sours. The conversion of ethanol into acetic acid is controlled and facilitated by an enzyme. The first step is for the enzyme, E, and the substrate (ethanol), S, to combine and form a complex written $E \cdot S$. In some cases, this complex falls apart before a product is formed, but in other cases, the enzyme catalyzes the reaction that leads to the substrate–enzyme complex forming a product (acetic acid), P, and the product quickly separates from the enzyme. This process can be summarized as

$$E + S \underset{k_{-1}}{\overset{k_1}{\rightleftharpoons}} E \cdot S \overset{k_2}{\rightarrow} E + P.$$

Reactions catalyzed by enzymes are central to much of biochemical engineering and bioprocess engineering. As you can learn in almost any course in

Chemical and Biomedical Engineering Calculations Using Python®, First Edition. Jeffrey J. Heys.
© 2017 John Wiley & Sons, Inc. Published 2017 by John Wiley & Sons, Inc.
Companion Website: www.wiley.com/go/heys/engineeringcalculations_python

those fields, if a process is controlled by the above set of reactions, then we can describe the change in the concentration of substrate in a close system (i.e., a system with no inflow and outflow like a sealed wine bottle) with the equation:

$$\frac{dS}{dt} = \frac{-V_{max}S}{K_m + S},$$

where S is the concentration of substrate, V_{max} is a parameter that describes the maximum reaction rate (i.e., full utilization of all enzymes because high concentration of S is present) and K_m is the substrate concentration at which the reaction rate is half of V_{max}. In order to solve an initial value problem, we seek to determine the substrate concentration, S, that satisfied the equation above, that is, the rate of change of the substrate is equal to a function that depends on the current concentration.

The equation for $\frac{dS}{dt}$ describes the loss or consumption of S. The rate at which S is consumed must be exactly equal to the rate at which P is produced (i.e., every ethanol molecule that is reacted away forms an acetic acid molecule) so we can also write an equation for the formation of P as

$$\frac{dP}{dt} = \frac{V_{max}S}{K_m + S}$$

(note the sign change). This model of enzyme kinetics is known as the Michaelis–Menten kinetics model, originally proposed in 1913! Given an initial concentration of substrate S and product P, we can solve the initial value problem to determine the change in these respective concentrations over time. Figure 11.1 shows the change in substrate and product concentration for an

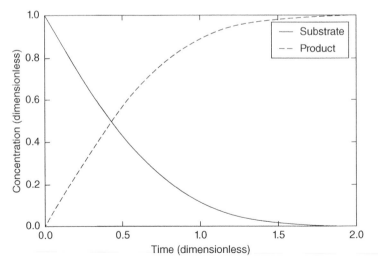

Figure 11.1 Substrate consumption (solid curve) and product formation (dashed curve) for a process governed by Michaelis–Menten kinetics.

initial concentration of 1.0 (dimensionless) for the substrate and a concentration of 0.0 for the product. The ordinary differential equations (ODEs) were solved using `scipy.integrate.odeint()`, which is described later in this chapter.

11.3 Forward Euler

In this section, the forward Euler method is described, and it is the simplest approach for solving an initial value problem. Any initial value problem can be written as

$$\frac{dy}{dt} = f(y, t), \tag{11.1}$$

and the initial condition $y_0 = y(t = 0)$ must be included. The key to understanding the process for approximately solving the initial value problem is to simply recognize that the initial value problem gives us (1) a starting point, y_0, and (2) a slope ($\frac{dy}{dt}$)! We can use these two pieces of information to make an estimate as to where the unknown function, y, is going in the near future. Taking a small step from the initial value in the direction given by the slope leads to a new value for y at a new time point that is slightly different from where we started. At this new location, we can again calculate the slope and take a small step in the direction given by the slope. This process is illustrated in Figure 11.2. The process described here estimates the future based ONLY on the current conditions. This category of time-stepping methods for solving initial value problems are called *explicit* methods because the estimates of the future are based explicitly on the present.

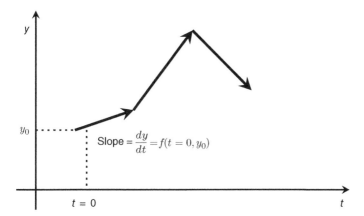

Figure 11.2 An approximate solution to an initial value problem can be obtained by calculating a slope from the ODE equation based on the current values for y and t, and then taking a small step in time to a new set of values. This problem is repeated until the desired final time is reached.

The construction of an algorithm that implements the process described above begins with the creation of a function that contains the ODE and returns the slope (i.e., the time derivative) based on a given time, t, and value for the dependent variable, y. The forward Euler process is based on an iteration where the slope is calculated and then, based on the size of the time step, δt, a new value for y is calculated. Recall from Chapter 10 that the first derivative of a function can be approximated with a finite difference approximation:

$$\frac{dy}{dt} \approx \frac{y_i - y_{i-1}}{\delta t} = f(y_{i-1}, t_{i-1}). \tag{11.2}$$

The current value of the dependent variable, y_{i-1}, and time, t_{i-1}, is used to calculate the slope and solve for the new value of the dependent variable, y_i. The previous equation can be rearranged to give

$$y_i = y_{i-1} + \delta t \cdot f(y_{i-1}, t_{i-1}), \tag{11.3}$$

which is the forward Euler method. The Python script below implements the forward Euler method to predict the change in substrate concentration for the Michaelis–Menten reaction.

```python
import numpy
import pylab

# model kinetic parameters
Vmax = 2.0 # mol/(L s)
Km = 0.5 # mol/L

# ODE definition
def df(s,t):
    dsdt = -Vmax*s/(Km+s)
    return dsdt

# setup time discretization
n = 10 # number of time steps
t = numpy.linspace(0,2.0,n)
dt = t[1]-t[0]

# allocate storage space and set initial conditions
sol = numpy.zeros(n)
sol[0] = 1.0 # initial S in mol/L

for i in range(1,n):
    sol[i] = sol[i-1]+dt*df(sol[i-1],t[i-1])

pylab.plot(t,sol)
pylab.xlabel("time (dimensionless)")
pylab.ylabel("concentration (dimensionless)")
pylab.show()
```

The first section of this script contains the ODE function within a separate, callable function. The inputs to this function are the current value for the dependent variable (substrate concentration) and the current time. The function returns the slope, $\frac{ds}{dt}$. The forward Euler algorithm requires that time be discretized into "small" segments, and, to facilitate this process, the algorithm builds a vector that holds all the time points using the `numpy.linspace()` function. The initial condition and storage space for the final solution are then set. The forward Euler iteration is very simple because it calculates a new value for the dependent variable based on the slope and the time step size.

Figure 11.3 shows a plot of the substrate concentration versus time using the forward Euler method with two different time step sizes. The solid curve is based on 100 time steps ($\delta t = 0.02$) and the dashed curve is based on 4 time steps ($\delta t = 0.5$). The large time steps used for the dashed curve show how the slope of the approximation is equal to the slope at the beginning of the time step. During the time step, the slope decreases as the substrate concentration is reduced so there is a significant error associated with using the large time step. The accuracy of the forward Euler method is of the same order as the forward finite difference approximation of the derivative, that is, the error is $O(\delta t)$. If the length of a time step is cut in half, the error is also halved. Fortunately, the forward Euler method has a relatively low computational cost so taking a large number of time steps is feasible here, but if we need to simulate a much longer period of time or if we are solving many initial value problems simultaneously, the very small time step requirement quickly becomes an issue. To begin to address this accuracy limitation, we now turn our attention to improving the order of accuracy.

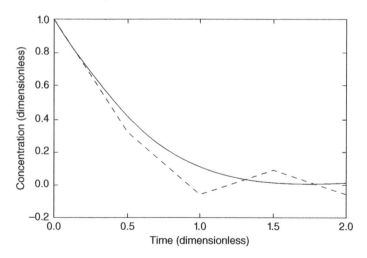

Figure 11.3 Substrate concentration governed by a Michaelis–Menten reaction is modeled using the forward Euler method with two different time step sizes: 100 time steps (solid) and 4 times steps (dashed).

11.4 Modified Euler Method

The forward Euler method is based on an estimate of the slope based only on the current conditions. One method for improving the accuracy of the forward Euler method is to predict future conditions and then use that prediction to get a better estimate for the slope. For example, we can predict future conditions using the forward Euler method:

$$y_i^* = y_{i-1} + \delta t \cdot f(y_{i-1}, t_{i-1}), \tag{11.4}$$

and then use this prediction to estimate the slope at t_i using

$$\frac{dy}{dt} = f(y_i^*, t_i).$$

We now have two estimates for the slope, one at t_{i-1} and one at t_i. Using an average of these two estimates gives us a better estimate of the slope over the time span of interest. Using this principle, the modified Euler method calculates a new value for the dependent variable using:

$$y_i = y_{i-1} + 0.5 \cdot \delta t \cdot (f(y_{i-1}, t_{i-1}) + f(y_i^*, t_i)), \tag{11.5}$$

where y_i^* is calculated using equation 11.4 (i.e., forward Euler). The accuracy of the modified Euler approach is $O(\delta t^2)$, which leads to significantly smaller errors for small time step sizes.

A Python script that utilizes the modified Euler method for the Michaelis–Menten kinetics problem is given below.

```python
import numpy
import pylab

# model kinetic parameters
Vmax = 2.0 # mol/(L s)
Km = 0.5 # mol/L

# ODE definition
def df(s,t):
    dsdt = -Vmax*s/(Km+s)
    return dsdt

# allocate space and set initial condition
n = 100 # number of time steps
# Vector of time points over the time domain
t = numpy.linspace(0,2.0,n)
# Create an empty vector for storing the solution
sol = numpy.zeros(n)
sol[0] = 1.0 # initial S in mol/L
# Calculate time step size
dt = t[1]-t[0]
```

```
# Iteration over time steps
for i in range(1,n):
    slope = df(sol[i-1],t[i-1])
    est = sol[i-1]+dt*slope
    sol[i] = sol[i-1]+0.5*dt*(slope+df(est,t[i]))

# Plot final solution
pylab.plot(t,sol)
pylab.xlabel("time (dimensionless)")
pylab.ylabel("concentration (dimensionless)")
pylab.show()
```

Overall, the algorithm is very similar to the forward Euler method, but it has an extra step in the iteration loop to estimate the value of the dependent variable at the end of the time step.

11.5 Systems of Equations

The enzymatic degradation of a substrate to a product involves changes in concentration to both the substrate and the product. In this and many other systems, a change in one system parameter has an impact on many other system parameters. For these problems, multiple initial value problems must be solved simultaneously. This implies that there is a vector, $\mathbf{y}$, of the dependent variable and a vector of the ODEs:

$$\frac{d\mathbf{y}}{dt} = \mathbf{f}(\mathbf{y}, t).$$

Fortunately, the extension of the methods described above to a system of ODEs is trivial. The ODE definition function needs to be modified slightly so that it can receive a vector of dependent variables and it must return a vector of derivatives (or slopes), but otherwise there is little change.

As we move toward systems of equations, the computational costs can increase quickly and it is often advantageous to utilize the initial value problem algorithms available in the scipy.integrate library. These algorithms have a number of helpful advantages:

1) high accuracy – often fourth order ($O(\delta t^4)$) or higher,
2) error checking – comparing the prediction of a fourth- and fifth-order method (or, in general, comparing the prediction of two methods with different orders of accuracy) each time step allows the algorithm to adjust the time step size in order to maintain a desired level of accuracy, and
3) fast execution – the algorithms are often written in FORTRAN or C and execute faster than a purely Python algorithm.

The simplest initial value problem solver in the Scipy library is scipy.integrate.odeint(). This function must be given the name of a function containing the ODEs, the initial condition(s), and the time span for integration. The use of this function for modeling both the substrate and product concentration for a Michaelis–Menten kinetics problem is illustrated in the Python script below.

```python
import numpy
import pylab
from scipy.integrate import odeint

# model kinetic parameters
Vmax = 2.0 # mol / (L s)
Km = 0.5 # mol / L

# ODE definition
def df(c,t):
    s = c[0] # substrate concentration
    p = c[1] # product concentration
    dsdt = -Vmax*s / (Km + s)
    dpdt = Vmax*s / (Km + s)
    return numpy.array([dsdt,dpdt])

# initial condition
c0 = numpy.array([1.0, 0.0]) # initial S, P in mol / L
t = numpy.linspace(0,2.0,100)
sol = odeint(df, c0, t)
pylab.plot(t,sol)
pylab.xlabel("time (dimensionless)")
pylab.ylabel("concentration (dimensionless)")
pylab.show()
```

Notice that a vector of dependent variables is passed into the function containing the ODEs. Care must be taken to ensure that *the order in which unknowns are located in the vector is the same order used for the derivatives that are returned.* Assigning new variable names to the unknowns in the dependent variable vector can be helpful for keeping track of the various unknowns. Of course, these new variable names are limited to the function and cannot be used outside the function. The figure resulting from the this Python script was shown earlier in this chapter (Figure 11.1).

11.5.1 The Lorenz System and Chaotic Solutions

In 1963, Edward Lorenz derived a mathematical model of atmospheric flows that consisted of a system of three initial value problems:

$$\frac{dx}{dt} = \sigma(y - x)$$

$$\frac{dy}{dt} = x(\rho - z) - y$$

$$\frac{dz}{dt} = x \cdot y - \beta z,$$

where Lorenz set $\sigma = 10$, $\beta = 8/3$, and $\rho = 28$. The Python script below solves this system of equations using the same parameters that Lorenz derived, and it uses random initial values for x, y, and z that are between -2 and $+2$.

```python
import numpy as np
import scipy.integrate as sint
import matplotlib.pyplot as plt
from mpl_toolkits.mplot3d import Axes3D

sigma = 10.0
beta = 8.0/3.0
rho = 28.0

def lorentz_deriv(xi, t):
    (x,y,z) = xi
    dxdt = sigma * (y - x)
    dydt = x * (rho - z) - y
    dzdt = x * y - beta * z
    return [dxdt, dydt, dzdt]

# Choose two random starting points, (-2, 2)
x0 = -2.0 + 4.0 * np.random.random(3)

# Solve the IVP
t = np.linspace(0, 4, 1000)
sol0 = sint.odeint(lorentz_deriv, x0, t)

# Plot the solution
fig = plt.figure()
ax = fig.gca(projection='3d')
ax.plot(sol0[:,0], sol0[:,1], sol0[:,2]
plt.savefig('LorenzFig0.png', dpi=150)
```

The script uses the same `scipy.integrate.odeint()` function that has been used previously in this chapter for solving initial value problems. The one unique feature of the script is that the `mpl_toolkits.mplot3d.Axes3D` library is used, and this library is connected with matplotlib and enables 3D plotting. The ability to generate plots with three axes is helpful here because we would like to plot the values of three different parameters (x, y, and z) at each time point.

The Lorenz equations are possibly the most famous initial value problem of the past 50 years, but they would be completely forgotten if it were not for one very interesting observation. Before the Lorenz equations were numerically solved on an early computer, the common assumption was that the solution to an initial value problem was not overly sensitive to the initial conditions used. Small changes in initial condition were assumed to be unlikely to cause a large change in the final solution. For example, Figure 11.4 shows two different solutions to the Lorenz system that have slightly different initial values. The two solution track close to each other and ultimate oscillate in very similar trajectories. This was the expected behavior for any two similar initial conditions.

Fortunately, when Lorenz entered the initial condition that he desired for his model into the computer, he made a mistake and changed the initial conditions slightly. The solution that he observed was completely different from what he expected. It took time for Lorenz to find his initial condition mistake, but, when he did, he recognized the implications of what he had found: *the solution to this system of equations was highly sensitive to the initial conditions*. Figure 11.5 shows an example of this behavior. The two initial conditions are similar, but the final solutions are very different. The implication of this result is critical: for some systems such as atmospheric dynamics (i.e., weather prediction), we must have highly accurate initial conditions because slight errors in initial conditions could lead to very inaccurate predictions. This observation led to the development of the field of chaos [1]. The fundamental behavior of chaotic systems is that they are unpredictable because we never know the initial conditions exactly. For example, if a butterfly flaps its wings off the coast of South America, and we do not know it, we might mispredict the weather in Florida in the coming weeks.

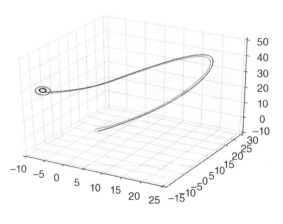

Figure 11.4 Two different solutions to the Lorenz system of equations using slightly different initial conditions. The three axes represent the three solutions variables, *x*, *y*, and *z*, and time is not shown other than the variables oscillate in a cycle long term.

Figure 11.5 Two different solutions to the Lorenz system of equations using slightly different initial conditions. The final solutions are very different even though the initial conditions are similar.

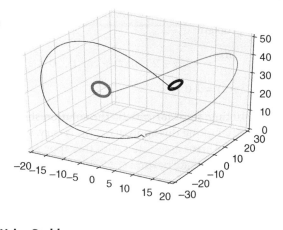

11.5.2   Second-Order Initial Value Problems

Although they are rare in some engineering fields, for example, chemical or biological engineering, some initial value problems involve second-order ODEs and two initial conditions. The classic problem in this category is the motion of a body under gravitational force. If x is the position of the body, then Newton's first law states:

$$\frac{d^2x}{dt^2} = -g, \tag{11.6}$$

where g is gravitational acceleration. Solving this equation requires two initial conditions: an initial position, $x(t = 0)$, and an initial velocity, $\frac{dx(t=0)}{dt}$.

Solving second-order initial value problems is relatively straightforward using the algorithms already described because second-order (or higher-order) problems can be rewritten as systems of first-order equations by defining new variables. For the problem above (equation 11.6), defining $v = \frac{dx}{dt}$ allows the second-order problem to be rewritten as a system of two first-order equations:

$$\frac{dx}{dt} = v, \tag{11.7}$$

$$\frac{dv}{dt} = -g, \tag{11.8}$$

and notice that one initial condition can be used for each equation: $x(t = 0)$ and $v(t = 0)$.

11.6   Stiff Differential Equations

The forward Euler method and other explicit time-stepping methods are very simple and computationally efficient for solving initial value problems, but they have an important limitation. Use the forward Euler method to solve the initial value problem:

$$\frac{dy}{dt} = -25y + 25\sin(t) + \cos(t), \quad 0 \le t \le 2.0$$
$$y(0) = 1.0$$

with 10 time steps ($\delta t = 0.2$). The solution shown in Figure 11.6 was obtained when attempting this approach. This is very, very far from the correct solution. In fact, the error associated with this approximation is growing exponentially, and if we integrate beyond $t = 2.0$, the approximation only becomes worse.

To help us understand why the forward Euler method failed so badly for this problem, it is helpful to look at the plot of the exact solution shown in Figure 11.7. The important thing to recognize about the solution ($y = \sin(t) + e^{-25t}$) is that there are two different time scales present – there is a very fast time scale that causes a rapid, initial decrease in the solution, and then there is a slower time scale associated with the oscillations from the sin() function. *The defining characteristic of stiff differential equations is two or more time scales.* If only the fast time scale existed, the solution would quickly reach steady state and the simulation could focus on the brief period when all the changes occur. If only the slow time scale existed, longer time steps could be used in obtaining an approximate solution. With stiff differential equations, the time step must be small enough to capture the fast time scale events, but those small time steps result in large computational cost associated with simulating the slower, longer time scale.

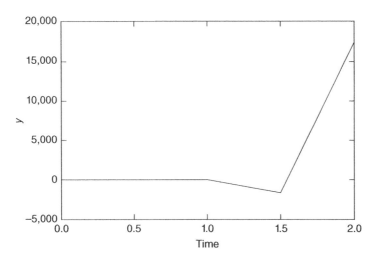

Figure 11.6 The approximate solution resulting from using the forward Euler method on the initial value problem $\frac{dy}{dt} = -25y + 25\sin(t) + \cos(t)$ with $y(0) = 1.0$ and $\delta t = 0.2$. The approximation error is large and growing exponentially.

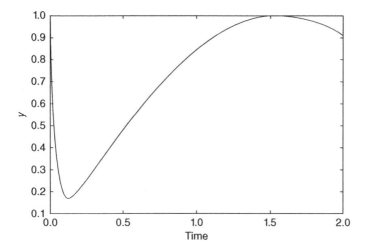

Figure 11.7 The exact solution and the solution resulting from using `scipy.integrate.odeint()` on the initial value problem $\frac{dy}{dt} = -25y + 25\sin(t) + \cos(t)$ with $y(0) = 1.0$ and $\delta t = 0.2$. The two curves overlap one another.

The use of explicit time-stepping methods on a stiff ODE reveals the major weakness of these methods. Recall that explicit methods predict the future direction of the solution using only currently available information. An analogy would be to walk around watching only the ground exactly at your feet – based on the topology at your feet, you take a step anticipating that the topology of the ground is not going to change dramatically over that step. For problems with only one time scale, that assumption holds. Stiff ODEs, however, have a slow changing topology (the slow time scale) and a cliff or fast changing topology. When an explicit method hits the fast time scale, it is analogous to stepping off a cliff and exponential error increases result.

A number of different algorithms have been developed to address the challenge of stiff ODEs. The algorithm used by `scipy.integrate.odeint()` is one such algorithm [2], and if this function is used to solve the stiff example problem above, the solution is indistinguishable from the exact solution. Figure 11.7 actually contains two curves – the exact solution and the approximate solution from `scipy.integrate.odeint()`. Algorithms designed for stiff ODEs utilize two strategies to handle the multiple time scales within the problem. The first strategy is to continuously check the accuracy of a predicted solution for each time step and then continuously adapt the size of the time step to the size required to maintain accuracy. By adjusting the time step size, these algorithms can take smaller time steps whenever faster time scales are causing rapid changes and take larger time steps whenever the changes are slow. This first strategy is helpful to minimize the computational costs associated with stiff ODEs, but, by itself, this strategy is not sufficient and a second, critical technique is required.

Explicit time-stepping methods for solving initial value problems all have the basic form:

$$y_i = y_{i-1} + F(\delta t, y_{i-1}), \tag{11.9}$$

where the predicted solution is only based on current values (or estimates of the future that are still, ultimately, based on current values as is the case of the modified Euler method or the popular, explicit Runge–Kutta methods.) The alternative to explicit time stepping is implicit time stepping where the slope or change in the dependent variables over the next time step is not simply based on current values but also based on future values. Implicit methods have the basic form:

$$y_i = y_{i-1} + F(\delta t, y_{i-1}, y_i). \tag{11.10}$$

For example, the simplest implicit method is the backward Euler method, and it has the form:

$$y_i = y_{i-1} + \delta t \cdot f(t_i, y_i) \tag{11.11}$$

for solving an initial value problem of the form: $\frac{dy}{dt} = f(t, y)$. Notice that the unknown variable, y_i, now appears in multiple places within the equation (11.11) and solving for the unknown likely requires solving a *nonlinear* equation. For multiple ODEs, we have to solve a system of nonlinear equations every time step. The development of algorithms that use implicit time stepping requires utilization of the methods covered in Chapter 8 on nonlinear equations – typically, these algorithms use Newton's method to solve the nonlinear equations. Fortunately, robust algorithms have been developed by others and should be utilized whenever possible. A number of different algorithms that should meet most requirements are available through the `scipy.integr ate.ode()` function (http://docs.scipy.org/doc/scipy/reference/generated/scipy.integrate.ode.html).

Problems

11.1 One of the most famous initial value problems is the predator–prey problem. If x is the population of prey, it is typical to assume that the population change is governed by

$$\frac{dx}{dt} = ax - bxy,$$

where a is the birthrate per unit of x and b the death rate due to the predator with population y. The population of predator is typically assumed to be governed by

$$\frac{dy}{dt} = cxy - dy,$$

where c is the growth rate from the consumption of prey and d is the death rate from overpopulation or age.

Solve the predator–prey model equations with $a = 2$, $b = 1$, $c = 1.5$, and $d = 2$, and with an initial population of 1.0 for both the predator and the prey. Plot the population over at least 10 time units.

11.2 Model the concentration of reactant A in a stirred tank reactor with two inputs. The only reaction is

$$A \rightarrow B.$$

This reaction can be described by the first-order reaction equation $-r_A = kC_A$ where $-r_A$ is the rate at which A is consumed in mol A/(L · s), C_A is the concentration of A in mol/L and $k = 0.35$ s^{-1} is the rate constant. There are two input streams into the reactor: the first input has a flow rate of $Q_1 = 10$ L/min and $C_{A,1} = 2$ mol A/L and the second input stream is turned on at $t = 0$ and has a flow rate of $Q_2 = 8$ L/min and a concentration of $C_{A,2} = 5$ mol A/L. The concentration of A in the tank is governed by the equation:

$$\frac{dC_A}{dt} = [C_{A,1}Q_1 + C_{A,2}Q_2 - (Q_1 + Q_2)C_A]/V - kC_A,$$

where $V = 50$ L is the volume of the reactor. Before $t = 0$, the reactor is operating at steady state, that is, $dC_A/dt = 0$, and $Q_2 = 0$ so the above equation simplifies to

$$(C_{A,1} - C_A)Q_1/V = kC_A,$$

which can be solved to establish that at $t = 0$, $C_A = 0.73$ mol/L. This is the initial condition that should be used to model the concentration of A in the reactor after $t = 0$. Write a Python script to model this process and plot C_A as a function of time.

11.3 You have been hired by the City Health Department to estimate the impact on an Ebola outbreak on a city of 50,000 people. You have been asked to use a relatively simple initial value problem model of Ebola. Let S be the number of healthy, but susceptible people leaving in the city (initial value for S is 50,000), let I be the number of people infected by Ebola (initial value for I is 2.0), and let R be the number of people that recovered from Ebola. The model equations are as follows:

$$\frac{dS}{dt} = -C \cdot S \cdot I$$

$$\frac{dI}{dt} = C \cdot S \cdot I - I - d \cdot I$$

$$\frac{dR}{dt} = I,$$

where C is the rate of contact between sick and healthy people (unknown, but estimates of C are 2.0/50,000 to 10.0/50,000) and d the rate of death relative to recovery (estimate to be 0.5, implying 1 person dies for every 2 that recover).

Solve this model of Ebola using `scipy.integrate.odeint()` for a model period of 30 days and for 3 different contact rates and write a report on the results to the City Health Department.

11.4 *Adapted from "Astronomy Projects for Calculus and Differential Equations" by Farshad Barman, Portland Community College, 2012.*

You have been hired by NASA to determine the minimum distance between Mars and Earth at any time in the next 10 years. Assuming a Cartesian coordinate system with the sun at $(0, 0)$ and both the planets orbiting in the (x, y)-plane, the location, (x, y), and velocity (v_x, v_y) of either planet is determined using the initial conditions and solving the following system of equations from Newton:

$$\frac{dx}{dt} = v_x$$

$$\frac{dy}{dt} = v_y$$

$$\frac{dv_x}{dt} = -\frac{G \cdot M \cdot x}{\left(\sqrt{x^2 + y^2}\right)^3}$$

$$\frac{dv_y}{dt} = -\frac{G \cdot M \cdot y}{\left(\sqrt{x^2 + y^2}\right)^3},$$

where G is the universal gravitational constant, $6.67 \times 10^{-11} \frac{m^3}{s^2 \cdot kg}$ or $1.983979 \times 10^{-29} \frac{AU^3}{year^2 \cdot kg}$, and M the mass of the sun, 2.0×10^{30} kg. Because of the large distance and time spans being simulated, it is recommended that the problem be solved using astronomical units (AU) as the units of distance and years as the units for time. There are 149.598×10^9 m/AU and 3.15569×10^7 s/year.

The initial conditions for the earth are as follows: $x(0) = 0.44503$ AU, $y(0) = 0.88106$ AU, $v_x(0) = -5.71113$ AU/year, and $v_y(0) = 2.80924$ AU/year. The initial conditions for the Mars are as follows: $x(0) = -0.81449$ AU, $y(0) = 1.41483$ AU, $v_x(0) = -4.23729$ AU/year, and $v_y(0) = -2.11473$ AU/year.

Your first project requirement is to solve the initial value problem for each planet and determine the location and velocity of each planet for the next 10 year period. NASA requests a plot of the orbits over that period. The second project requirement is to loop through each time point from the 10-year solutions and determine the minimum distance between the two planets at any time during that period. NASA recommends obtaining the solution for 10,000 time points during the 10-year period. Report the minimum distance in AU.

11.5 Halocarbons are molecules containing carbon and at least one atom of chlorine, bromine, or iodine. The most common halocarbons are chlorofluorocarbons (CFCs), which were commonly used gases up until the late 1980s. CFCs are relatively stable, long-lived molecules that, regardless of their purpose, inevitably escape into the troposphere (the lowest layer of the earth's atmosphere). The compounds then cycle between the troposphere and the stratosphere. While in the stratosphere, a portion of the molecular compound (the chlorine and bromine atoms) can be dissociated, and these atoms catalyze the removal of ozone. The atoms remain in the stratosphere on average 3 years before they are transported back to the troposphere where they are removed by rain or surface deposition. A mathematical model of this process was published by Ko *et al.* [3]. The model has one mass balance on an initial quantity of halocarbons in the troposphere:

$$\frac{dH_T}{dt} = -\frac{H_T}{L_T} - \frac{H_T \cdot f}{\tau} + \frac{H_S}{\tau},$$

where H_T and H_S are the amount of halocarbon in the troposphere and the stratosphere, respectively. The first term on the right represents halocarbons that chemically degrade, the second term is halocarbons transported from the troposphere into the stratosphere, and the third term is halocarbons that return to the troposphere from the stratosphere. The time scales for these events, based on experimental measurements, are $L_T = 1000$ years and $\tau = 3$ years. The second mass balance is on the quantity of halocarbons in the stratosphere:

$$\frac{dH_S}{dt} = -\frac{H_S}{L_S} - \frac{H_S}{\tau} + \frac{H_T \cdot f}{\tau},$$

where $L_S = 5$ years because hydrocarbons are disassociated more rapidly in the stratosphere. The final balance is on the quantity of free chlorine in the stratosphere:

$$\frac{dC}{dt} = -\frac{C}{\tau} + \frac{H_S}{L_S}.$$

You have been hired as a consultant to independently solve these equations using a numerical solver. You should assume that 15% of the atmospheric mass is in the stratosphere so $f = 0.15/0.18$. The initial conditions that you should use are $H_T = 1\,\text{kg}$ and $H_S = C = 0.0$. The key result will be a plot of the concentration of each of the three species as a function of time. You should simulate a sufficient time period that all three quantities are practically zero (maybe 100 years?).

Bonus: you are offered a bonus if you can integrate the C concentration over that 100-year period using the numerical integration approaches from the previous chapter. The total amount of ozone loss due to this 1 kg of CFCs depends on $\int C \, dt$.

References

1 Gleick, J. (1987) *Chaos: Making a New Science*, Open Road Integrated Media, New York.
2 Hindmarsh, A. (1983) ODEPACK, a systematized collection of ODE solvers, in *Scientific Computing* (ed. R. Stepleman), Elsevier, North-Holland, Amsterdam.
3 Ko, M., Sze, N.D., and Prather, M. (1994) Better protection of the ozone layer. *Nature*, **367**, 505–508.

12

Boundary Value Problems

This chapter continues our exploration of numerical methods to solve ordinary differential equations (ODEs), which have derivatives with respect to a single independent variable. The focus here is on problems that have a second-order derivative so two conditions are required to determine a unique solution. In the previous chapter, the derivatives were generally with respect to time, and the second-order problems had two initial conditions – one on the dependent variable and the other on the first derivative of the dependent variable. In this chapter, the derivatives will typically be with respect to space, and the two conditions used to determine a unique solution will be at either end of the spatial domain that is being modeled.

12.1   Introduction

Boundary value problems (BVPs) frequently arise in engineering. A general form for a linear BVP equation is

$$\frac{d^2y}{dx^2} + a(x)\frac{dy}{dx} + b(x) \cdot y = c(x). \tag{12.1}$$

Solving a BVP requires finding a function, $y(x)$, that satisfies this equation (i.e., find y such that the second derivative of y plus $a(x)$ times the first derivative plus $b(x)$ times y is equal to a given right-hand side). Determining a unique solution, that is, determining y requires two additional conditions on y, and without those boundary conditions, there are an infinite range of possible functions $y(x)$ that solve the ODE equation. Another way to see that we need two boundary conditions is to recognize that we need to integrate this equation twice, which gives two constants of integration, and we need the two additional conditions to solve for the two constants of integration. For some problems, $y(x)$ is known at one or both ends of the domain. This boundary condition is referred to as a Dirichlet or essential boundary conditions. For example, if we are solving the BVP above on the domain [0, 1] and it is known that $y(0) = 3.2$, this would be considered a Dirichlet condition. If Dirichlet conditions are given

Chemical and Biomedical Engineering Calculations Using Python®, First Edition. Jeffrey J. Heys.
© 2017 John Wiley & Sons, Inc. Published 2017 by John Wiley & Sons, Inc.
Companion Website: www.wiley.com/go/heys/engineeringcalculations_python

for both ends of the domain, that is, $y(1)$ is also known, then an approximate solution can be determined. The other common possibility is that $\frac{dy}{dx}$ is known at one or both ends of the domain. This type of boundary condition is known as a Neumann boundary condition.

BVPs arise when describing diffusion processes, when modeling conductive heat transport, and when calculating the velocity of viscous fluid flow. There are many other settings in which these types of equations can potentially arise, but problems involving conservation of mass, energy, and momentum are the most common in engineering. In this chapter, two different numerical approaches will be examined for solving BVPs – the shooting method and the finite difference method. In both the cases, we will utilize numerical techniques that were covered previously, so be prepared to review material from early chapters as needed.

12.2   Shooting Method

The previous chapter discussed an approach for calculating the location of a projectile under just the force of gravity given two initial conditions: an initial location and an initial velocity (recall that velocity is just the first derivative of location, $\frac{dx}{dt}$). The equation describing the motion of the projectile was a second-order ODE, but the problem fell into the category of initial value problems because it included two conditions at the same boundary (i.e., the $t = 0$ boundary) instead of one condition at each boundary. BVPs have a condition at each boundary, analogous to solving the projectile motion problem given a starting location and an ending location but no initial velocity. Without two initial conditions, we cannot simply reuse the methods covered in the previous chapter, but, with a little creativity, we can recycle much of what was developed previously and adapt it to BVPs.

Imaging that we are solving the projectile motion problem, given an initial location and a target location, and we need to determine an initial velocity as well as the particle location and velocity between the launching and target locations. We could guess an initial velocity, take a shot, and determine the final distance from the given target. We could then take a second shot with a different initial velocity, and once again measure the distance to the target. Using these two shots as reference points, we could then interpolate (or extrapolate) to determine a better estimate for the required initial velocity to hit the target. Repeating this process for three or four shots would hopefully lead to us hitting the target. This process is effectively "the shooting method" for BVPs. We guess one initial value, use all the initial value methods from the last chapter, check to see if we matched the other boundary condition, and repeat the process until our guess at the unknown initial value results in us matching the second boundary condition.

It is easiest to examine the algorithm for the shooting method through an example problem. We want to solve the BVP:

$$\frac{d^2y}{dx^2} = 4(y - x) \tag{12.2}$$

on the domain, $0 \le x \le 1$, and with the boundary conditions, $y(0) = 0$ and $y(1) = 2$. The first step is to rewrite equation 12.2 as a system of first-order equations by introducing a new variable:

$$\frac{dy}{dx} = y_1 \tag{12.3}$$

$$\frac{dy_1}{dx} = 4(y - x), \tag{12.4}$$

where an initial condition is available for the first equation ($y(0) = 0$) but not the second equation, that is, $y_1(0)$ is unknown. The shooting method requires a guess for the second boundary condition, and then using the guess, we can solve the system of equations (12.3 and 12.4) using any of the algorithm presented in the initial value problem chapter (the `scipy.integrate.odeint()` function is recommended) and check to see if the other boundary condition ($y(1) = 2$) is satisfied. Figure 12.1 shows the solution using a guess of $y_1(0) = 1.0$. With this guess, the target of 2.0 is missed and $y(1) = 1.0$ is hit instead.

For the second shot, an initial guess of $y_1(0) = 0.0$ is used, and the approximate solution using this guess is shown in Figure 12.2. In this case, the target is missed by a greater amount as $y(1) = -0.813$.

Now that two shots have been taken and the two misses have been measured, we need a method for determining a better guess for the second boundary

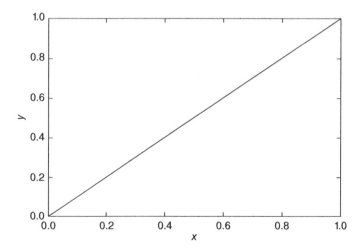

Figure 12.1 Approximate solution to the BPV, $\frac{d^2y}{dx^2} = 4(y - x)$ using a guess of $\frac{dy(0)}{dx} = y_1(0) = 1.0$.

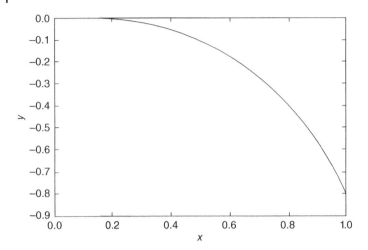

Figure 12.2 Approximate solution to the BPV, $\frac{d^2y}{dx^2} = 4(y - x)$ using a guess of $\frac{dy(0)}{dx} = y_1(0) = 0.0$.

condition on $y_1(0)$. The simplest approach is to fit the two previous results with a line and extrapolate to determine a better guess. If γ_0 is the first guess at the boundary condition, γ_1 is the second guess, and β is the desired target value, then an improved guess is available using:

$$\gamma = \gamma_1 - \frac{\gamma_1 - \gamma_0}{(y_{\gamma_1}(1.0) - y_{\gamma_0}(1.0))}(y_{\gamma_1} - \beta). \tag{12.5}$$

An iterative process is used where the improved estimate for the boundary condition, γ, replaces the older guess, γ_0, and the process is repeated.

A Python script that uses the shooting method to solve the example problem is given below.

```python
import math
import numpy
import pylab
from scipy.integrate import odeint

# Split y'' = 4*(y-t) into
# y0' = y1 and
# y1' = 4*(y-t)
def dfdt(y, t):
    dy0dt = y[1]
    dy1dt = 4.0*(y[0]-t)
    return numpy.array([dy0dt, dy1dt])
```

```python
def exact(t):
    coeff = 0.13786
    sol=coeff*(numpy.exp(2.0*t)-numpy.exp(-2.0*t))+t
    return sol

TOL = 1e-6
t = numpy.linspace(0.0,1.0,100)
alpha = 0.0
beta = 2.0
gamma0 = 1.0
gamma1 = 0.0

# first shot, use bc for y, set other to 0.0
yinit1 = numpy.array([alpha,gamma0])
y1 = odeint(dfdt, yinit1, t)
# get impact point for first shot
# note this gets the last row, first column entry
end1 = y1[-1,0]
print("Error with shot: ", math.fabs(beta-end1))

for i in range(20):
    # second shot, set bc for y to 0.0, other uses 1.0
    yinit2 = numpy.array([alpha,gamma1])
    y2 = odeint(dfdt, yinit2, t)
    end2 = y2[-1,0]
    print("Error with shot: ", math.fabs(beta-end2))
    if math.fabs(beta-end2) < TOL:
        break

    gamma = gamma1
    gamma -= (end2-beta)*(gamma1-gamma0)/(end2-end1)
    gamma0 = gamma1
    gamma1 = gamma
    end1 = end2

pylab.plot(t,y2[:,0])
pylab.plot(t,exact(t))
pylab.xlabel('x')
pylab.ylabel('y')
pylab.show()
```

The Python script includes an additional function that contains the exact solution so that the approximate solution from the shooting method can be compared to the exact solution. One feature of numpy that is used in this script but has not been covered previously is that it is possible to access the last element in a row or column of an array using the index −1. For example, the reference $y[-1]$ in numpy will return the last entry in the vector. This feature is used here to get the last row entry in the first column of the solution array to determine the value of y at the far boundary, that is, $y(1)$ so that we

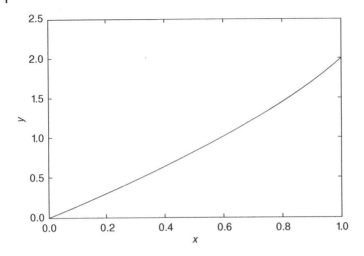

Figure 12.3 Approximate solution to the BPV, $\frac{d^2y}{dx^2} = 4(y - x)$ using a guess of $\frac{dy(0)}{dx} = y_1(0) = 1.551$.

can calculate the distance to the target, β. Figure 12.3 shows both the final result from the shooting method and the exact solution. The lines are so close that they are indistinguishable from each other. For linear BVPs like this, only three iterations should be required because the linear extrapolation used to determine an improved guess, γ, should yield the exact value to use. Nonlinear BVPs may require four or five iterations to determine an acceptable value for γ.

The shooting method is a common choice for BVPs when an efficient and familiar ODE initial value problem solver is available and Dirichlet boundary conditions are given. In general, however, it is not the most common choice for BVPs in the author's experience. A more intuitive and flexible approach, capable of solving problems with one or two Neumann boundary conditions, is presented in the following section (and the next chapter): the finite difference method.

12.3 Finite Difference Method

The finite difference method is based on the idea of replacing the derivatives in a differential equation with algebraic approximations of those derivatives at discrete points distributed throughout the domain of interest. A general form for a linear BVP is

$$\frac{d^2y}{dx^2} + a(x)\frac{dy}{dx} + b(x) \cdot y = c(x), \tag{12.6}$$

on the domain $a \leq x \leq b$ with boundary condition given at a and b. The finite difference method begins by dividing the domain, $a \leq x \leq b$, into a sequence

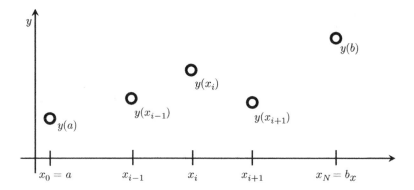

Figure 12.4 Using the finite difference method requires dividing the domain, $a \leq x \leq b$ into a set of discrete points or nodes with their locations given by x_i. The goal of the approach is to determine the approximate solution, y_i, at every node.

of evenly spaced, discrete points called nodes with their location given by x_i, as shown in Figure 12.4. If h is the distance between the nodes, $h = x_i - x_{i-1}$, and N is the number of intervals between nodes (i.e., there are $N + 1$ nodes that are numbered from 0 to N), then the location of each node can be calculated using $x_i = a + i \cdot h$.

Recall (equation 10.12) that the second derivative can be approximated at location x_i with

$$\frac{d^2y}{dx^2} \approx \frac{y_{i+1} - 2y_i + y_{i-1}}{h^2}, \tag{12.7}$$

where h is the distance between the nodes, $h = x_i - x_{i-1}$, and y_i is short for the value of y at node x_i or $y(x_i)$. Similarly, the first derivative (equation 10.8) can be approximated with

$$\frac{dy}{dx} \approx \frac{y_{i+1} - y_{i-1}}{2h}. \tag{12.8}$$

Using these two approximations for the derivatives, the original BVP equation can be replaced with an algebraic approximation at every node in the domain (i.e., we are replacing a differential equation with $N + 1$ algebraic equations). The algebraic approximation is

$$\left(\frac{y_{i+1} - 2y_i + y_{i-1}}{h^2} \right) + a(x_i) \left(\frac{y_{i+1} - y_{i-1}}{2h} \right) + b(x_i)y_i = c(x_i) \tag{12.9}$$

Notice that we have transformed the problem from a differential equation into a large system of linear algebraic equations. The unknowns are the values of y_i at every node.

The finite difference algorithm for linear BVPs has three sections:

1) The setup phase involves specifying the number of intervals between nodes, N, which translates into $N + 1$ nodes numbered from 0 to N, the size of the

domain by specifying a and b, and calculating the node spacing, h. The setup phase also typically includes building a vector containing the locations of the nodes, and allocating space for later storing the matrix and right-hand side associated with the linear system of equations.

2) The middle section of the code is a loop through each node and adding the appropriate coefficients into the matrix and right-hand side. The details of this step are summarized below, but it is important to note that nodes located at the boundary have boundary conditions that need to be handled separately.

3) The last section of the algorithm involves solving the linear matrix problem (using, e.g., `numpy.linalg.solve()`) and plotting the solution.

The finite difference equation at every node has the form:

$$\left(\frac{y_{i+1} - 2y_i + y_{i-1}}{h^2}\right) + a(x_i)\left(\frac{y_{i+1} - y_{i-1}}{2h}\right) + b(x_i)y_i = c(x_i),$$

which is typically rewritten as

$$(y_{i+1} - 2y_i + y_{i-1}) + \frac{h}{2} \cdot a(x_i)(y_{i+1} - y_{i-1}) + h^2 b(x_i)y_i = h^2 c(x_i). \quad (12.10)$$

This equation exists at every node, and the resulting system of equations can be written as a matrix problem with the following form:

$$\begin{bmatrix} -2 + h^2 b(x_1) & 1.0 + \frac{ha(x_1)}{2} & 0 & \cdots \\ 1.0 - \frac{ha(x_2)}{2} & -2 + h^2 b(x_2) & 1.0 + \frac{ha(x_2)}{2} & 0 \\ 0 & 1.0 - \frac{ha(x_3)}{2} & -2 + h^2 b(x_3) & \ddots \\ \vdots & 0 & 1.0 - \frac{ha(x_4)}{2} & \ddots \end{bmatrix} \cdot \begin{bmatrix} y_1 \\ y_2 \\ \vdots \\ y_N \end{bmatrix}$$

$$= \begin{bmatrix} h^2 c(x_1) \\ h^2 c(x_2) \\ \vdots \\ h^2 c(x_N) \end{bmatrix} \quad (12.11)$$

It is important to note that the above linear matrix system does NOT include boundary conditions (or an equation for node 0) and is only intended to give the basic structure of the linear system. If the boundary condition $y(x_0) = 1.0$ is given, then instead of having a finite difference equation for the first node, which corresponds to the first row in the matrix equation, the boundary condition equation would be used instead and the matrix problem would become

$$
\begin{bmatrix}
1.0 & 0 & 0 & \cdots \\
1.0 - \frac{ha(x_1)}{2} & -2 + h^2 b(x_1) & 1.0 + \frac{ha(x_1)}{2} & 0 \\
0 & 1.0 - \frac{ha(x_2)}{2} & -2 + h^2 b(x_2) & \ddots \\
\vdots & 0 & 1.0 - \frac{ha(x_3)}{2} & \ddots
\end{bmatrix}
\begin{bmatrix}
y_0 \\
y_1 \\
\vdots \\
y_N
\end{bmatrix}
$$

$$
=
\begin{bmatrix}
1.0 \\
h^2 c(x_1) \\
\vdots \\
h^2 c(x_N)
\end{bmatrix}
\tag{12.12}
$$

The finite difference method is used below to solve the BVP:

$$
\frac{d^2 y}{dx^2} + y = 0
$$

on the domain $0 \le x \le \pi/2$ with the Dirichlet boundary conditions $x(0) = 1.0$ and $x(\pi/2) = 1.0$. The finite difference equation at each node for this equation is

$$
(y_{i+1} - 2y_i + y_{i-1}) + h^2 \cdot y_i = 0.0.
\tag{12.13}
$$

The Python script is below.

```python
import numpy
from numpy.linalg import solve
import pylab

N = 9 # number of intervals
x = numpy.linspace(0,numpy.pi/2.0,N+1)
h = x[1]-x[0]

# Allocate space
A=numpy.zeros((N+1,N+1))
b = numpy.zeros(N+1)

# Boundary condition at x=0
A[0,0] = 1.0
b[0] = 1.0

for i in range(1,N):
    A[i,i-1] = 1.0
    A[i,i] = -2.0 + h**2
    A[i,i+1] = 1.0
    b[i] = 0
```

```
# Boundary condition at x = pi/2.0
A[N,N] = 1.0
b[N] = 1.0

y = solve(A,b)
pylab.plot(x,y)

pylab.plot(x, numpy.cos(x)+numpy.sin(x))
pylab.xlabel('x')
pylab.ylabel('y')
pylab.show()
```

The equations associated with the boundary conditions (i.e., the equations associated with nodes at the boundary) receive special handling in this algorithm, but all the other finite difference equations (12.13) are handled with an iterative loop. For this particular problem, an exact solution is available and this solution is plotted on the same figure as the approximate solution. Even with only 10 nodes, the finite difference approximation is almost indistinguishable from the exact solution as shown in Figure 12.5.

12.3.1 Reactions in Spherical Catalysts

Many chemical and biochemical reactions ($A \rightarrow B$) are facilitated by catalysts, and these catalysts are often solid particles that the reactants diffuse into until the reaction occurs at a catalytic site. Microbes that are used in biological

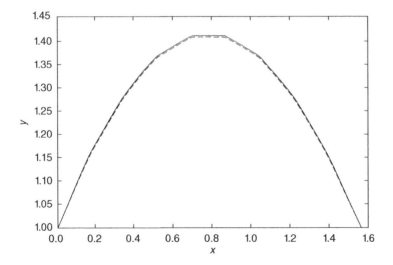

Figure 12.5 Approximate solution to the BPV, $\frac{d^2y}{dx^2} + y = 0$ using the finite difference method with 10 nodes. The exact solution is also plotted (dashed line) and is almost indistinguishable from the approximate solution.

processes to facilitate various reactions are often found in flocs, which are biofilm particles that are similar to the solid catalyst particles of traditional chemical processes. In either scenario, the physical processes of interest include (1) diffusion into the catalyst particle and (2) reaction within the particle. A material balance on a spherical catalyst particle gives the following equation that includes both physical processes:

$$\frac{\partial^2 c}{\partial r^2} + \frac{2}{r}\frac{\partial c}{\partial r} - \phi^2 c = 0, \tag{12.14}$$

where r is the radial distance from the center of the sphere, c is the concentration of the reactant, and ϕ is the Thiele modulus, a dimensionless parameter that describes the relationship between the reaction rate and the diffusion rate. The two left most terms in the material balance capture diffusion, and the term with the Thiele modulus captures the reaction. If the Thiele modulus is large, then the reaction is much faster than diffusion and most of the reaction happens near the surface and the concentration of reactant is zero over most of the particle because the reactant is consumed before it can diffuse into the particle. If the Thiele modulus is small, then the reaction is slow and the particle has a high concentration of reactant everywhere.

The first derivative term in equation 12.14 must also be approximated using a finite difference approximation. Using the standard centered difference approximation, the first-order derivative term will be replaced with

$$\frac{2}{r}\frac{\partial c}{\partial r} \approx \frac{1}{r}\frac{c_{i+1} - c_{i-1}}{h}.$$

Note that the 2 in the numerator cancels with the 2 that appears in the denominator of the centered difference approximation (equation 10.8). As a result of this additional first derivative term, the terms off the main diagonal of our finite difference matrix are going to have the form:

$$1.0 - \frac{h}{r} \tag{12.15}$$

after multiplying each equation by h^2. It may be helpful to compare the off-diagonal term in equation 12.15 with that in equation 12.12, and note that the $a(x)$ term from earlier is set to $\frac{2}{r}$ here. Further, the terms along the main diagonal are going to have the form

$$-2.0 - \phi^2 \cdot h^2$$

due to the reaction term.

Typical boundary conditions for the reaction in a spherical catalyst equation 12.14 are

$$\frac{dc}{dr} = 0 \text{ at } r = 0,$$

which basically says that if the concentration is the same on all surfaces of the catalyst, then the concentration has to be symmetric inside the catalyst. This boundary condition can be rewritten using a finite difference approximation for the first derivative as

$$\frac{c_1 - c_0}{h} = 0.0,$$

where c_0 is the concentration at the center $(r = 0)$ and c_1 is the concentration at the point nearest the center $(r = h)$. This equation can be simplified to $c_0 = c_1$, and this equation will be used as the first equation in the linear system of equations. The second boundary condition is that the concentration must be known on the surface of the catalyst particle. For example, if the bulk concentration of the reactant is 2.0 (dimensionless), then the boundary condition is

$$c = 2.0 \text{ at } r = 1.0,$$

assuming that the radius of the particle is 1.0 (nondimensionalized).

The governing equation 12.14 and boundary conditions are solved using the finite difference method in the script below.

```
import numpy
from numpy.linalg import solve
import matplotlib.pyplot as plt

N = 99 # number of intervals
phi = 5

# finite difference point spacing
r = numpy.linspace(0,1.0,N+1)
h = r[1]-r[0]

# Allocate space
A = numpy.zeros((N+1,N+1))
b = numpy.zeros(N+1)

# Symmetry boundary condition at r=0
A[0,0] = -1.0
A[0,1] = 1.0
b[0] = 0.0

for i in range(1,N):
    A[i,i-1] = 1.0 - h/r[i]
    A[i,i] = -2.0 - phi**2 * h**2
    A[i,i+1] = 1.0 + h/r[i]
    b[i] = 0.0

# Concentration boundary condition at r = 1.0
A[N,N] = 1.0
b[N] = 2.0

c = solve(A,b)
```

```
plt.plot(r,c)
plt.xlabel('radius')
plt.ylabel('concentration')
plt.title('$\phi = 5$')
```

The approximate solution to equation 12.14 is shown in Figure 12.6 for two different values of ϕ. Note that the left edge of the figures shows the predicted concentration at the center of the particle and the right edge of the figures shows the concentration at the surface, which should always be 2.0. Figure 12.6(a) shows $\phi = 1.0$, which represents a reaction rate that is approximately equal to the rate of diffusion, and Figure 12.6(b) shows $\phi = 5$, which represents a reaction rate that is 5 times the rate of diffusion causing much lower concentrations near the center of the spherical catalyst particle. The Python script above uses the fact that the matplotlib library supports advanced mathematical equation editing when placing a title on a plot. The "$" symbols are used to indicate a mathematical equation, and everything appearing between the "$" signs is processed using LaTeX. One particularly

Figure 12.6 Concentration inside a spherical catalyst where a first-order reaction is occurring. The center of the particle is $r = 0$, and the surface is at $r = 1$. The concentration at the surface is $c = 2.0$. (a) $\phi = 1$ and (b) $\phi = 5$, which represents a much faster reaction rate.

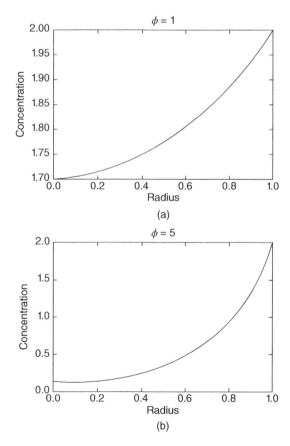

helpful feature that is used above is that Greek letters can be obtained by placing a \ in front of the name of the letter. In the Python script above, \phi is converted into the Greek letter, ϕ. For more information regarding using LaTeX to format mathematical equations, see [1] or any other LaTeX reference.

Problems

12.1 Use the shooting method to solve the BVP:

$$\frac{d^2y}{dx^2} = -y$$

on the domain $0 \le x \le \frac{\pi}{4}$ with $y(0) = 1$ and $y(\pi/4) = 1.0$.

12.2 Use the finite difference method to solve the BVP:

$$\frac{d^2y}{dx^2} = -\cos(x) - \sin(x)$$

on the domain $0 \le x \le \pi/2$ and with the boundary conditions $y(0) = 0$ and $\frac{dy(\pi/2)}{dx} = -1.0$. Note that this problem has the solution $y = \cos(x) + \sin(x)$.

The one complication with this problem is the Neumann boundary condition at $x = \pi/2$. The derivative in this boundary condition can be approximated as

$$\frac{y_{x_N} - y_{x_{N-1}}}{h} = -1.0.$$

This equation becomes the last row in the linear matrix problem, which has the form:

$$\begin{bmatrix} \ddots & 1.0 + \frac{ha(x_{N-2})}{2} & 0 \\ 1.0 - \frac{ha(x_{N-1})}{2} & -2 + h^2b(x_{N-1}) & 1.0 + \frac{ha(x_{N-1})}{2} \\ 0 & -1.0 & 1.0 \end{bmatrix} \cdot \begin{bmatrix} y_{N-2} \\ y_{N-1} \\ y_N \end{bmatrix}$$

$$= \begin{bmatrix} 0.0 \\ 0.0 \\ -h \end{bmatrix}.$$

12.3 You have been hired by Fisser Pharmaceuticals to analyze the potential of a new drug for treating vitreous hemorrhage in the eye. The new drug would be delivered topically to the cornea via eye drops. The drug would diffuse toward the back of the eye (i.e., to the retina), and any drug reaching the retina in the back of the eye would immediately be taken away by the blood flow in the retina. The challenge is that the drug naturally decays relatively quickly as it diffuses away from the cornea toward the retina.

The diffusion and decay of the drug are described by the equation:

$$\frac{d^2 c}{dx^2} - k \cdot c = 0,$$

where the first term captures diffusion and the second term is decay. The boundary conditions are $c = 1.0$ (dimensionless) at $x = 0.0$ (the cornea) and $c = 0.0$ (dimensionless) at $x = 2.0$ cm (the retina). For the current version of the drug, the decay constant is $k = 20$ cm^{-2}.

The first question Fisser would like you to answer is: What is the concentration at the vitreous hemorrhage site, which is at the center of the vitreous at $x = 1.0$ cm? Is the concentration greater than 0.2 (dimensionless), the minimum effective concentration?

The second question is what decay rate, k, would approximately give a concentration of 0.2 (dimensionless) at $x = 1.0$ cm?

12.4 You have been hired as a consultant by a company that manufactures tubing for blood donations. The tubing transports blood from the needle at the injection site to the collection bag. The inner wall of the tube is 1 cm from the center of the tube (i.e., the inner diameter of the tube is 2 cm) and is at blood or body temperature (38 °C). The outer wall of the tube is at room temperature (23 °C) and is 1.3 or 1.8 cm from the center of the tube (Figure 12.7). The temperature of the tubing wall material varies between the inner wall temperature and the outer wall temperature and is governed by the equation:

$$\frac{d^2 T}{dr^2} + \frac{1}{r} \frac{dT}{dr} = 0.$$

The approximate solution to this equation can be found by dividing the domain, $1.0 \leq r \leq 1.3$ or $1.0 \leq r \leq 1.8$ into discrete points and solving the finite difference approximation equation at each point:

$$\left(\frac{T_{i+1} - 2T_i + T_{i-1}}{h^2} \right) + \frac{1}{r_i} \left(\frac{T_{i+1} - T_{i-1}}{2 \cdot h} \right) = 0,$$

where h is the distance between the discrete points in the radial direction, T_i the temperature of point i, and r_i the radial location of point i.

The company is debating between the tubing with an outer wall diameter of 1.3 and 1.8 cm. One factor in choosing the best diameter tube is that there is some concern that large temperature gradients (approximated as $(T_i - T_{i-1})/h$) will cause the tubing to fracture and fail. Your consulting contract requires you to submit a brief report with your recommendation regarding tubing thickness selection. The report should summarize your findings and include figures showing temperature versus radius for

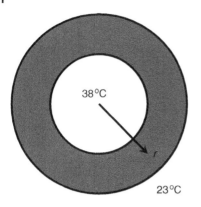

Figure 12.7 A thick cylindrical tube for transporting blood from an individual into a collection bag. The inner wall (1.0 cm from the center) of the tube is at 38 °C and the outer wall (1.3–1.8 cm from the center) is at 23 °C.

the two different tube thicknesses and reports the temperature gradient at some location for each tube thickness (Figure 12.7).

Reference

1 Lamport, L. (1994) *LaTeX: A Document Preparation System*, Addison-Wesley, Upper Saddle River, NJ, 2nd edn.

13

Partial Differential Equations

A goal in developing a mathematical model is to create the simplest possible model that captures the features of interest in the system. In many cases, the variables we are interested in calculating are primarily changing in time, and spatial variation can be either ignored or captured through a lumping parameter. For example, in a well-mixed reactor or even a region of the body, it is often possible to develop useful and informative models that include only time derivatives and not spatial derivatives. These models are typically initial value problems. In other cases, there are important spatial variations in temperature or concentration, but because the system is continuous and nearly steady state, the small temporal variations can be ignored. For the case of a membrane or a large slab, it is often sufficient to only model spatial changes in one direction, which leads to a one-dimensional boundary value problem. However, in other cases, it is necessary to examine variation of the quantity of interest in multiple dimensions – either time and spatial variation or variation in multiple spatial dimensions, which leads to partial differential equations (PDEs). These equations have two or more independent variables and they have derivatives with respect to each of these variables.

13.1   Finite Difference Method for Steady-State PDEs

The finite difference method was developed in the previous chapter for the solution of boundary value problems. The objective of this chapter is to extend this method to multiple dimensions and PDEs. This category of problem is still called boundary value problems, but algorithm development is much more complex and fraught with pitfalls when multiple dimensions are involved. Finite difference algorithms for PDEs are more complex than any algorithm previously developed in this book. The presentation here will break the full algorithm into its major sections and each section will be discussed independently. The primary example used in this chapter is the solution to

Chemical and Biomedical Engineering Calculations Using Python®, First Edition. Jeffrey J. Heys.
© 2017 John Wiley & Sons, Inc. Published 2017 by John Wiley & Sons, Inc.
Companion Website: www.wiley.com/go/heys/engineeringcalculations_python

Laplace's equation in two dimensions:

$$\frac{\partial^2 c}{\partial x^2} + \frac{\partial^2 c}{\partial y^2} = f(x, y). \tag{13.1}$$

Solving this equation requires the determination of the function c such that its second derivative with respect to x plus its second derivative with respect to y is equal to $f(x, y)$, and c must satisfy the required boundary conditions. This equation describes diffusion through solids and biological tissues, and it describes conductive heat transport through a stationary medium. At the end of this section, a second example is presented that includes a convective flux term.

13.1.1 Setup

Finite difference algorithms typically begin with a setup phase where the size of the domain and the number of nodes are specified. The algorithm developed here is based on M intervals (i.e., $M + 1$ nodes) in the x-direction and N intervals in the y-direction. Ideally, the spacing between the nodes in each direction is equal but that is not required. What is required is that the spacing between the nodes in each direction be *uniform*. Figure 13.1 shows a typical node in the domain. If i is used as an index for nodes in the x-direction, and j is used as an index for nodes in the y-direction, then the center node in this figure is $node(i, j)$. The four nearest neighboring nodes are also shown in this figure. As we will see below, each node is effectively coupled to these four nearest neighbors when we use the finite difference approximation presented later.

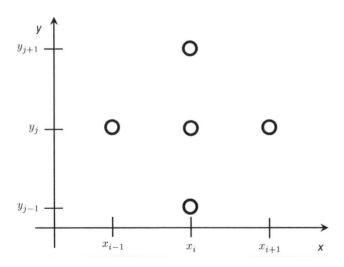

Figure 13.1 If the finite difference method is used for a two-dimensional boundary value problem, each node is connected to its four nearest neighbors.

The first section of Python code for the two-dimensional Laplace problem using finite differences is shown below.

```python
import numpy
from numpy.linalg import solve
import pylab

M = 15 # intervals in x-direction
N = 15 # intervals in y-direction

west = 0.0 # edge locations
east = 1.0
south = 0.0
north = 1.0

# node locations and spacing
x = numpy.linspace(west,east,M+1)
y = numpy.linspace(south,north,N+1)
h = x[1]-x[0]
k = y[1] - y[0]
h2k2 = h**2 / k**2
coeff = 2*(h2k2+1)

# plot the mesh/grid
X,Y = numpy.meshgrid(x,y)
pylab.plot(X,Y,'o')

# Allocate space
totalNodes = (M+1)*(N+1)
A=numpy.zeros((totalNodes,totalNodes))
b = numpy.zeros(totalNodes)
```

The `scipy.linalg.solve()` function is imported because it is used later for solving the linear matrix problem that is the result of the finite difference approximation. The variables M and N are set to the number of intervals between nodes in the x and y directions, respectively. Larger values for M and N will give a more accurate approximate solution, but values larger than about 100 will have a significant computational cost. The total number of nodes in the domain is $(M + 1)(N + 1)$ so $M = 100$ and $N = 100$ will result in over 10,000 nodes and a linear matrix problem with over 10,000 unknowns. This is about the largest acceptable matrix size for a solver based on Gaussian elimination. The variables `east, west, south, north` are used to specify the location of the boundaries of the rectangular domain of the problem. Some authors of numerical algorithms find those variable names to be intuitive while others dislike (hate) those variable names because they cannot remember which direction is east and which is west, for example. Different variable names are, of course, acceptable. For example, some algorithm writers use the variables left, right, bottom, top instead or even a, b, c, d.

The next stage in the setup phase of this algorithm is to build vectors holding the locations of the nodes in the x and y directions. On the basis of these node locations, the spacing between the nodes, h and k, can be calculated. The variables h2k2 and coeff are needed later in the algorithm. The numpy.meshgrid() function converts the vectors x and y into 2D arrays that hold the x and y locations of every node. These arrays are only needed for plotting the solution at the end of the algorithm although they can be used here to generate a plot with the locations of the nodes included (this line is currently commented out). The final step in the setup phase is to calculate the total number of nodes and, hence, the total number of unknowns in the linear matrix problem. Once this is known, space for the matrix and right-hand-side vector can be allocated.

13.1.2 Matrix Assembly

The model problem that is used here is

$$\frac{\partial^2 c}{\partial x^2} + \frac{\partial^2 c}{\partial y^2} = 0 \qquad (13.2)$$

on the domain $0 \le x \le 1.0$ and $0 \le y \le 1.0$ and $c = 0$ on all boundaries except for the west (left) boundary, which has $c = \sin(y)$. Approximating the derivatives with centered finite difference equations (10.12) leads to the algebraic equation

$$\frac{c_{i+1,j} - 2c_{i,j} + c_{i-1,j}}{h^2} + \frac{c_{i,j+1} - 2c_{i,j} + c_{i,j-1}}{k^2} = 0 \qquad (13.3)$$

and multiplying this equation by h^2 leads to

$$-2\left[\left(\frac{h}{k}\right)^2 + 1\right] c_{i,j} + c_{i+1,j} + c_{i-1,j} + \left(\frac{h}{k}\right)^2 (c_{i,j+1} + c_{i,j-1}) =$$

$$h^2 f(x,y) = 0. \qquad (13.4)$$

This algebraic, finite difference approximation of the original PDE exists at each node and makes up one row (or one equation) in the linear matrix problem. Equation 13.4 includes coefficients for five unknowns, reflecting that each equation connects a center node, $c_{i,j}$, to its four nearest neighbors (north, east, south, and west).

Assembly of the linear matrix problem begins with an outer loop through each row of nodes in the y-direction followed by an inner loop through each column of nodes in the x-direction. The combination of these two loops is that each node in the domain is visited once – starting in the lower left corner (southwest corner), proceeding to the right across the first row, then up to the next row, and then next until we end in the upper right corner (northeast corner). At each node, the appropriate coefficients from equation 13.4 are added to the matrix, unless the node is located on a boundary and then the correct boundary condition must be added to the linear system of equations.

Before examining the algorithm for assembling the matrix, it is important to recognize that we need a mechanism for mapping node (i, j) to a unique row in the matrix (i.e., to a unique unknown or node number). In the previous chapter, where we examined one-dimensional problems, this was trivial because node i corresponded to the ith unknown and ith row of the matrix. Now, however, we have node (i, j) due to the two-dimensional nature of the problem. (At this point, some students suggest using a three-dimensional matrix, but this is not the correct solution.) In order to uniquely map node (i, j) to a particular unknown number, we will use the equation

$$node = j * (M + 1) + i, \tag{13.5}$$

where $i = (0, \ldots, M)$ and $j = (0, \ldots, N)$. Therefore, for node $(0, 0)$, we get $node = 0$; for node $(0, 1)$, we get $node = M + 1$; and for node (M, N), we get $node = N * (M + 1) + M = (N + 1) * (M + 1) - 1$. This same equation can be used to calculate the unique node number of the neighboring nodes to the north, east, south, and west.

The section of Python code below assembles the matrix for the test problem.

```
for j in range(0,N+1):
    for i in range(0,M+1):
        node = j*(M+1)+i
        Enode = j*(M+1)+i+1 # node to the east
        Wnode = j*(M+1)+i-1 # node to the west
        Snode = (j-1)*(M+1)+i # node to the south
        Nnode = (j+1)*(M+1)+i # node to the north

        if (i == 0): # check for west boundary
            A[node,node] = 1.0
            b[node] = numpy.sin(numpy.pi*y[j])
        elif (i == M): # check for east boundary
            A[node,node] = 1.0
            b[node] = 0.0
        elif (j == 0): # check for south boundary
            A[node,node] = 1.0
            b[node] = 0.0
        elif (j == N): # check for north boundary
            A[node,node] = 1.0
            b[node] = 0.0
```

```
else:
    A[node,node]  = -coeff
    A[node,Enode] = 1.0
    A[node,Wnode] = 1.0
    A[node,Snode] = h2k2
    A[node,Nnode] = h2k2
    b[node] = h**2 * 0.0
```

This section of code consists of two nested loops that cause us to ultimately loop through every node in the domain. The numbers of the node and its neighbors are calculated first. Then, the algorithm checks to see if a node is on the boundary. It is important to do this first because if a node is on the boundary, then one or two of its neighbors does not exist and trying to write entries into the matrix for nodes that do not exist will only lead to crashes and error messages. If the node is on the boundary, the appropriate boundary condition can be applied (in this case, all boundary conditions are Dirichlet conditions and all are zero except for the east boundary). Finally, if the node is not on the boundary, the appropriate values, based on the finite difference equation 13.4 , are added into the matrix and the right-hand side.

13.1.3 Solving and Plotting

The final code segment solves the linear matrix problem (`scipy.linalg.solve()` is recommended), and the result is returned as a one-dimensional vector. The contour plotting function in Matplotlib typically requires that the solution data be in an array with the same shape as the arrays (X and Y) that hold the locations of the nodes for the two-dimensional finite difference mesh. Each row in the array corresponds to a row of nodes in the domain. The solution vector can be reshaped using the `numpy.resphape()` function before the plotting function is called. Depending on the plotting routine that is used, additional labels and color bars may be helpful. The Python code segment below finalizes the process of solving the Laplace problem in two dimensions using finite differences. The resulting contour plot is shown in Figure 13.2

```
z = solve(A,b)
Z = z.reshape(M+1,N+1)

# Plotting
CT = pylab.contour(X,Y,Z)
pylab.clabel(CT)
pylab.xlabel('x')
pylab.ylabel('y')
pylab.show()
```

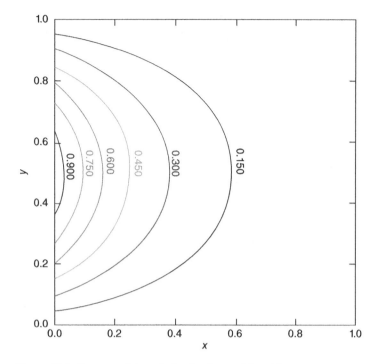

Figure 13.2 Contours of the solution for the model problem.

13.2 Convection

The Laplace problem examined in the first part of this chapter is used to model diffusion or heat conduction in stagnant domains. If there is also fluid movement or convection in addition to the diffusion or conduction, then an additional term is required in the model equation. To model steady-state diffusion and convection in two dimensions with the fluid velocity being given by the vector (u, v), the following equation is used:

$$\frac{\partial^2 c}{\partial x^2} + \frac{\partial^2 c}{\partial y^2} - u \cdot \frac{\partial c}{\partial x} - v \cdot \frac{\partial c}{\partial y} = 0. \tag{13.6}$$

Using the same boundary conditions and domain as before, and approximating the second-order derivatives with centered finite difference equations (10.12) and the first-order derivatives with backward finite difference equations (10.7) leads to the algebraic equation

$$\frac{c_{i+1,j} - 2c_{i,j} + c_{i-1,j}}{h^2} + \frac{c_{i,j+1} - 2c_{i,j} + c_{i,j-1}}{k^2} -$$
$$u \cdot \frac{c_{i,j} - c_{i-1,j}}{h} - v \cdot \frac{c_{i,j} - c_{i,j-1}}{k} = 0. \tag{13.7}$$

It is very important to recognize that less accurate, backward difference equations were used here instead of the more accurate centered difference equations (10.8) for first-order derivatives. When applying a finite difference approximation to the convective term, it is important that the approximation of the derivative be done in such a way as to include only the nodes that are upwind of the node of interest. The mathematical justification for this choice is available in a number of excellent books [1], but here we will simply state that not using an upwind difference approximation usually leads to numerical instability and significant error in the approximate solution. An analogy of questionable accuracy is that it is difficult to detect changes in smell (concentration) when facing downwind but it is relatively simple when facing upwind. The finite difference equation above is based on the assumption that both components of the wind velocity, (u, v), are positive. If one or both of these velocities is negative, then the finite difference equation must be modified to use the forward first-derivative approximations (equation 10.5) in the upwind direction.

Adding convection to the previous two-dimensional finite difference code is relatively simple and only requires (1) adding a wind velocity variable and (2) adding additional terms to the matrix for the convective part of the equation. The Python script for the test problem with a fluid velocity vector of $(5.0, 5.0)$ is reproduced completely below.

```python
import numpy
from numpy.linalg import solve
import pylab

M = 15 # intervals in x-direction
N = 15 # intervals in y-direction

west = 0.0 # edge locations
east = 1.0
south = 0.0
north = 1.0

# node locations
x = numpy.linspace(west,east,M+1)
y = numpy.linspace(south,north,N+1)
h = x[1]-x[0]
k = y[1] - y[0]
h2k2 = h**2 / k**2
coeff = 2*(h2k2+1)

# plotting the grid/mesh
X,Y = numpy.meshgrid(x,y)
```

```
pylab.plot(X,Y,'o')

# Wind
# both MUST be positive due to differencing
wind = numpy.array([5.0,5.0])

# Allocate space
totalNodes = (M+1)*(N+1)
A=numpy.zeros((totalNodes,totalNodes))
b = numpy.zeros(totalNodes)

for j in range(0,N+1):
    for i in range(0,M+1):
        node = j*(M+1)+i
        Enode = j*(M+1)+i+1 # node to the east
        Wnode = j*(M+1)+i-1 # node to the west
        Nnode = (j+1)*(M+1)+i # node to the north
        Snode = (j-1)*(M+1)+i # node to the south

        if (i == 0): # check for west boundary
            A[node,node] = 1.0
            b[node] = numpy.sin(numpy.pi*y[j])
        elif (i == M): # check for east boundary
            A[node,node] = 1.0
            b[node] = 0.0
        elif (j == 0): # check for south boundary
            A[node,node] = 1.0
            b[node] = 0.0
        elif (j == N): # check for north boundary
            A[node,node] = 1.0
            b[node] = 0.0
        else:
            A[node,node] = \
            -coeff - h*wind[0] - h**2*wind[1]/k
            A[node,Enode] = 1.0
            A[node,Wnode] = 1.0 + h*wind[0]
            A[node,Nnode] = h2k2
            A[node,Snode] = h2k2 + h**2*wind[1]/k
            b[node] = h**2 * 0.0

z = solve(A,b)
Z = z.reshape(M+1,N+1)

# Plotting solution
CT = pylab.contour(X,Y,Z)
```

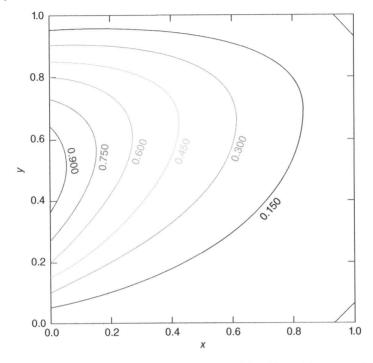

Figure 13.3 Contours of the solution for the model problem with a convective flow of (5.0, 5.0), that is, toward the northeast corner.

```
pylab.clabel(CT)
pylab.xlabel('x')
pylab.ylabel('y')
pylab.show()
```

The convective term can have a significant impact on the solution to the model problem, depending on the magnitude of the wind. The contours associated with a fluid velocity of (5.0, 5.0) are shown in Figure 13.3. Once the magnitude of the fluid velocity exceeds about 100, the problem quickly becomes more numerically demanding and a much finer mesh (i.e., more nodes) is likely to be required to get an accurate result. An inaccurate result is almost always inexpensive from a computational standpoint, but accuracy can be costly.

13.3 Finite Difference Method for Transient PDEs

PDEs have multiple independent variables, and the first part of this chapter examined problems where all the independent variables were spatial variables.

These equations described changes in two or more spatial directions. The other possibility is that one of the independent variables is time. The final part of this chapter examines problems where there are derivatives with respect to time and space in the same equation. This class of PDEs is often called "parabolic" PDEs. Before examining a finite difference algorithm to solve a parabolic PDE problem, let us briefly discuss controlled drug release.

Researchers are increasingly developing devices that gradually release a pharmaceutical drug over time. In most cases, a gradual release is preferred to the burst release that is associated with the injection of a drug. One type of device is based on the embedding of the drug in a polymer and allowing the drug to gradually diffuse out of the polymer over time. The polymeric material is eventually passed through the digestive track after the drug has been gradually released via diffusion. If we assume that the device is shaped like a chip (think poker chip or coin), then the diffusion occurs primarily in one spatial direction – the direction that requires the shortest distance for diffusion. In this case, the concentration is described by the equation

$$\frac{dc}{dt} = D\frac{d^2c}{dx^2}, \tag{13.8}$$

where D is the diffusivity and c the concentration within the device. We assume that the distance from the surface of the device, $x = 0$, to the center is 1.0 (dimensionless). As a result, the spatial domain is $0 \leq x \leq 1.0$. Further, we assume that the concentration at the surface is zero, that is, $c(x = 0) = 0.0$, implying the any drug that diffuses to the surface is immediately swept away, and the device is assumed symmetric about the center, that is,

$$\frac{dc}{dx}\bigg|_{x=0} = 0.$$

Finally, we assume that the initial concentration is 1.0 (dimensionless) everywhere except the surface, that is, $c(t = 0, 0 < x \leq 1.0) = 1.0$.

We have previously discussed how to replace the derivatives in the PDE with finite difference approximations. For this problem, the time that we wish to simulate is going to be divided into N discrete steps and we are going to approximate the time derivative with a forward difference approximation (i.e., use the forward Euler method, equation 11.3). The one spatial dimension is going to be divided into M intervals (or $M + 1$ nodes) that span from 0.0 to 1.0, and the second derivative in space will be replaced with a centered difference approximation (10.12). The result of the finite difference approximation is that the original PDE is replaced with the finite difference equation:

$$\frac{c_{i,new} - c_{i,old}}{\delta t} = D\frac{c_{i+1,old} - 2c_{i,old} + c_{i-1,old}}{h^2}, \tag{13.9}$$

where i is the index for the spatial location of the node, $c_{i,old}$ refers to the concentration at node i at the previous time step, and $c_{i,new}$ refers to the concentration at the next time step (i.e., the unknowns). This equation is usually rearranged to give

$$c_{i,new} = c_{i,old} + \delta t \cdot \frac{D}{h^2}[c_{i+1,old} - 2c_{i,old} + c_{i-1,old}]. \tag{13.10}$$

Notice that explicit time stepping is used so we can calculate the new concentration using only concentrations from the previous time step. This equation also shows how the new concentration at each node depends on three concentrations from the previous time step – the concentration at the same node and the two neighboring nodes. Figure 13.4 summarizes the connection between the concentration at a node and its neighboring nodes from the previous time step.

A Python script that utilizes the finite difference approximation to solve the transient diffusion problem, that is, the drug release problem, is shown below. The algorithm is similar to the others shown in this chapter in that the first section of the algorithm performs some basic setup operations. The number of time and space intervals is specified, the size of the domain and duration of the simulation are set, and some vectors containing the spatial node locations and time points are constructed. One additional and important step in the setup phase is that a vector must be constructed that contains the initial conditions.

```python
import numpy
import pylab

M = 15 # intervals in x-direction
N = 200 # total number of time steps
```

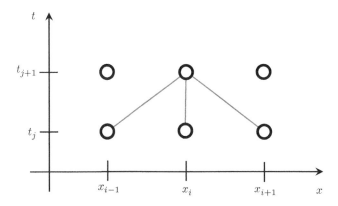

Figure 13.4 Diagram illustrating the finite difference approximation for a transient PDE with one spatial dimension. Time is shown on the *y*-axis and space is on the *x*-axis. When solving for the concentration at a node for the next time step, the concentration will depend on three adjacent nodes from the previous time step.

```
left = 0.0 # edge locations
right = 1.0
start = 0.0
stop = 20.0
diffh2 = 5.0

# node locations
x = numpy.linspace(left,right,M+1)
t = numpy.linspace(start,stop,N+1)
h = x[1]-x[0]
dt = t[1] - t[0]

# Initial conditions
c_old = numpy.ones(M+1)
c_old[0] = 0.0
c_new = numpy.zeros(M+1)
# Plot initial conditions
pylab.plot(x,c_old)
pylab.xlabel('x')
pylab.ylabel('c')

# time loop
for j in range(1,N+1):
    c_new[0] = 0.0
# spatial loop
    for i in range(1,M):
        tmp = c_old[i-1]-2*c_old[i]+c_old[i+1]
        c_new[i] = c_old[i]+dt*diffh2*(tmp)
    c_new[M] = c_new[M-1]
    c_old = c_new.copy()
# Plot every 20th time step
    if j%20 == 0:
        pylab.plot(x,c_new)

pylab.show()
```

The main section of the algorithm consists of two loops. The outer loop is a loop through the time steps, and the inner loop cycles through the spatial nodes. The algorithm operates by calculating new concentrations at every spatial location before moving on to the next time step. Care must be taken to set the boundary conditions. The Dirichlet boundary condition at $x = 0$ is set before proceeding through the inner loop through the spatial nodes. The Neumann boundary condition at $x = 1$ is enforced by setting the concentration of the edge node to be equal to its nearest neighbor. The justification for doing this comes from the boundary condition equation:

$$\frac{dc}{dx}\bigg|_{x=1} = \frac{c_M - c_{M-1}}{h} = 0$$

or

$$c_M = c_{M-1}.$$

This simulation uses a large number of time steps (200) and plotting the solution at every time step creates a very busy figure. To simplify the figure showing the results, the solution is only plotted every 20th time step by checking to see if the remainder of dividing the time step number by 20 is zero. The results from running the algorithm are shown in Figure 13.5.

Observant readers may be surprised that the algorithm above used 200 time steps and only 16 spatial nodes. This choice can be justified in part by observing that the approximation of the time derivative is first-order (temporal error is $O(\delta t)$ and the spatial derivative approximation is second-order $(O(h^2))$ so we should expect to need a higher temporal resolution. An alternative justification can be provided by simply rerunning the algorithm with only 40 time steps. The result of this numerical experiment is shown in Figure 13.6. In this case, the approximation error is seen to grown exponentially – a clear symptom of a stiff differential equation. Recalling that stiff differential equations should be solved using an implicit time-stepping method, we simply state here that switching to an implicit time-stepping method like backward Euler will avoid the exponential error growth seen here. The disadvantage of switching to implicit time stepping is that a linear (or nonlinear) matrix problem must be solved every time step. For this reason, many algorithms use a sufficiently small step size

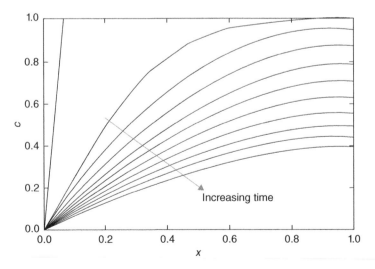

Figure 13.5 Plot of the concentration for the model problem shown every 20th time step. The initial concentration is the highest, and as time increases, the concentration decreases.

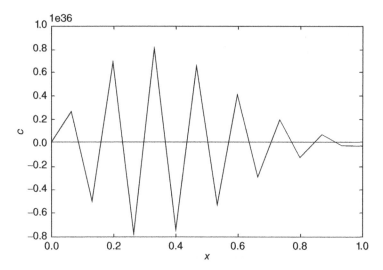

Figure 13.6 Plot of the concentration for the model problem using only 40 time steps. The error grows exponentially, and the final solution is completely without value.

to maintain stability. The time step size must be kept smaller than a constant times the node spacing squared, that is, $\delta t < kh^2$. Techniques for determining k are available in most numerical analysis books or k can be determined through numerical experiments.

Problems

13.1 A few days every year, it is possible to detect an unpleasant smell on the campus of the University of Colorado at Boulder. The (incorrect) explanation given to students was that this was the "Husker smell" emanating from Nebraska. (The correct explanation identified a company in Greeley, Colorado, that occasionally produced bad smells.) Develop a model of convection and diffusion over the state of Colorado. Model Colorado as a unit square (feel free to be more accurate if you are so compelled as the state is definitely not a square) and apply zero concentration boundary conditions to all boundaries except the eastern half of the north boundary (i.e., the right half of the top boundary) and the northern half of the east boundary (i.e., the top half of the right boundary), which roughly corresponds to the shared border between Nebraska and Colorado. (It may be helpful to consult a map of the United States.) Along the portion of Colorado's border that is shared with Nebraska, apply a concentration of 1.0. Determine the concentration in the center of the state of Colorado on a windless day and then the concentration for wind velocities in the x-direction only between -10.0 and $+10.0$.

13.2   You have been hired by the Department of Homeland Security to study the potential for Canada to attack Montana with a noxious odor. The concern is that Canada will release a noxious odor from either Lethbridge or Medicine Hat (both towns are just north of the northern border of Montana) and the noxious odor will diffuse in all directions, including into Montana. The odor will drive people out of Montana and then the Canadian Mounties (see, Do-Right, Dudly) will be able to move in and occupy Montana.

The diffusion of the noxious odor is governed by the two-dimensional Laplace equation:

$$\frac{\partial^2 c}{\partial x^2} + \frac{\partial^2 c}{\partial y^2} = 0,$$

where the domain of interest is the state of Montana, which should be approximated as a rectangle that is 500 miles west-to-east and 250 miles south-to-north. If the noxious odor is released from Medicine Hat, then the concentration of the odor along the north border should be approximated as

$$c = \sin\left(\pi \cdot \frac{x}{500}\right)$$

because Medicine Hat is north of central Montana. The concentration along all other borders (west, south, and east) should be approximated as zero. If the noxious order is release from Lethbridge, which is north of western Montana, then the concentration along the north border should be approximated:

$$c = \sin\left(\pi \cdot \frac{x}{250}\right), 0 \leq x \leq 250$$

and

$$c = 0, 250 < x \leq 500,$$

where x is always the distance from the western border of Montana.

In both the cases, the peak concentration along the border is 1.0 (dimensionless). You should plot the concentration everywhere in Montana and, in particular, determine the concentration in Bozeman, Montana, which can be approximated as 100 miles from the western border and 50 miles from the southern border. The citizens of Bozeman will be able to stay and defend the state if the concentration is less than 0.10 (dimensionless).

The final concern from the Department of Homeland is that the noxious odor will be release during a polar vortex, which will generate a wind from North to South with a velocity of -2.0 (dimensionless). In this case, the transport of the noxious odor will be governed by the equation:

$$\frac{\partial^2 c}{\partial x^2} + \frac{\partial^2 c}{\partial y^2} = -2.0 \cdot \frac{\partial c}{\partial y}.$$

Repeat the previous analysis under conditions of the polar vortex.

13.3 Nicotine patches are placed on the skin of an individual and they diffusively deliver nicotine across the skin and other tissues. They are primarily used by individuals that are trying to stop consuming cigarettes and provide an alternative nicotine source during the transition period away from cigarettes. You have been hired by a manufacturer of nicotine patches to predict the concentration of nicotine in the tissue near the patch over a 24- h time period.

The concentration of nicotine in the tissue can be accurately described by the equation:

$$\frac{\partial c}{\partial t} = D\frac{\partial^2 c}{\partial x^2} - k \cdot c,$$

where c is the concentration in mg/cm^3, t the time in h, $D = 0.1$ cm^2/h the diffusivity, and k the first-order decay rate of nicotine in h^{-1}.

The company has asked you to approximately solve this equation using an explicit finite difference approximation, using the finite difference equation:

$$c_{i,new} = c_{i,old} + \delta t \cdot \frac{D}{h^2}[c_{i+1,old} - 2c_{i,old} + c_{i-1,old}] - \delta t \cdot k \cdot c_{i,old}$$

with an initial concentration of zero everywhere except the skin surface adjacent to the patch. The skin surface next to the patch should maintain a fixed concentration of 1.0 mg/cm^3 for at least the 24-h period you have been asked to model. The soft tissue beneath the patch is 2-cm thick and then an impermeable bone prevents further diffusion of the nicotine. As a result, the model should have a no-flux boundary condition (i.e., $c_{i_{max}} = c_{i_{max}-1}$) at the other end of the spatial domain, 2 cm away from the skin surface.

The rate at which nicotine decays in the tissue varies significantly from one individual to another. At one extreme, some individuals lack the proteins that normally break down the nicotine so k is effectively zero. At the other extreme, some individuals rapidly break down nicotine so k can be as high as 1.0 h^{-1}.

Your report to the company should contain predictions (i.e., figures) of nicotine concentration at different points between the skin and the bone at different time points for a total time period of 24 h. The predictions should include both extreme cases, $k = 0.0$ h^{-1} and $k = 1.0$ h^{-1}. It is also suggested that you vary the number of spatial points/intervals and the number of time steps to convince yourself and the company that has hired you that the approximate solution is a good approximation to the original equation (hint: a good approximation should change little with changes in the number of temporal or spatial intervals).

Reference

1 Donea, J. and Huerta, A. (2003) *Finite Element Methods for Flow Problems*, John Wiley & Sons, Ltd., West Sussex, England.

14

Finite Element Method

This is an optional chapter on the use of the finite element method (FEM) and the FEniCS library. The FEniCS library can be installed in two different ways:

1) Windows and MacOS uses should begin by installing the Docker Toolbox (www.docker.com/products/docker-toolbox), and then following the instructions on the FEniCS project download page (fenicsproject.org/ download). This approach will install virtual machine software on Windows or MacOS that will run the Linux operating system within Windows or MacOS. The biggest disadvantage of this approach is that it can be difficult to share files between the base Windows or MacOS operating system and the Linux operating system in the virtual machine. The FEniCS Docker page is very helpful for setting up Docker and FEniCS properly with Windows (fenics-containers.readthedocs.io/en/latest).
2) If you have a computer running Ubuntu Linux (or one of the derivatives of this operating system such as Mint), you can add the PPA for FEniCS and install FEniCS with a single command (`sudo apt-get install fenics`).

As of 2016, the FEniCS library required Python 2.7 and support did not yet exist for Python 3.x.

14.1 A Warning

The FEM is an advanced numerical approach to solving partial differential equations and is typically only taught at the graduate level. This chapter only briefly develops the FEM. The interested reader is referred to the suggested reading at the end of the chapter for further information regarding FEM. The primary focus here is to present a tool (FEniCS) that advanced readers may appreciate later in their academic careers for solving PDEs in more complicated geometries and with greater computational efficiency.

Chemical and Biomedical Engineering Calculations Using Python®, First Edition. Jeffrey J. Heys.
© 2017 John Wiley & Sons, Inc. Published 2017 by John Wiley & Sons, Inc.
Companion Website: www.wiley.com/go/heys/engineeringcalculations_python

14.2   Why FEM?

The FEM is used to solve partial differential equation such as those that were solved earlier using the finite difference method. Why, we might ask, should we discuss anything beyond the finite difference method? Why do people use FEM instead of finite differences? There are a number of reasons for using FEM, but three reasons are typically given more often than any others:

1) The FEM is easy to use when the shape of the domain is complex. If we wish to solve Laplace's equation on a domain that is shaped like the human brain (and this is a real problem in medical imaging), then we need to use FEM because finite difference cannot accurately capture the shape of complex domains.
2) The FEM is typically easier to extend to higher order approximations. Implementing a fourth- or even eighth-order accurate FEM method is relatively straightforward. This level of accuracy is rarely needed, but it can be a real advantage in some situations like modeling blood flow in the aorta that is nearly turbulent.
3) Rigorous mathematical analysis of FEM is much more extensive than analysis of finite differences.

14.3   Laplace's Equation

The use of FEniCS and the FEM will only be briefly introduced here through the lens of an example problem. An excellent book [1] and a large library of example problems (fenicsproject.org) are available for FEniCS, and the interested reader should utilize these resources to gain a much more comprehensive understanding of the FEniCS library and FEM. The FEniCS project is actually a collect of software packages that are used in concert to solve differential equations using FEM. The central software package that to some extent ties everything together is called dolfin. Many of the packages, including dolfin, are written in C++ for faster execution, but wherever necessary, the C++ code has been wrapped so that it can be called from Python. To utilize the various FEniCS packages within a Python code, the first step is to import the dolfin library with **from** dolfin **import** *. Since this is often the only library that is imported, we do not need to worry about importing multiple objects with the same name.

14.3.1   The Mesh

Whenever we use FEM, we always need to start with a mesh. A mesh is a division of the problem domain into small polyhedral shapes. In two dimensions, we typically divide the domain into triangles or quadrilateral shapes. For simple domains, such as squares, rectangles, and circles, FEniCS has built in mesh generators that automatically divide the domain into triangles. The function call

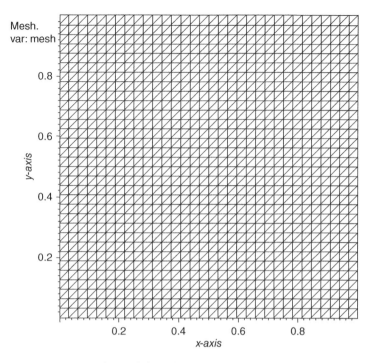

Figure 14.1 Triangular mesh from the `UnitSquareMesh()` function.

`mesh = UnitSquareMesh(32, 32)` divides a unit square into triangles with 32 triangles in each direction. The resulting mesh is shown in Figure 14.1. For more complex shapes, a number of software packages are available that provide a CAD-like interface for defining the domain (e.g., our domain might be a mountain bike) and then divide the domain into small polyhedral shapes of a desired type and size. The Cubit software from Sandia National Laboratory is one example.

14.3.2 Discretization

The basic idea behind the FEM is to select a set of mathematical functions called the function space and then determine the exact function on each element that best satisfies the original PDE. The most common function space by a significant margin is polynomial functions of a selected order. While this is a pretty simple idea, the actual implementation is much more difficult. To construct a function space in FEniCS, call `V = FunctionSpace(mesh, 'Lagrange', 2)` where "Lagrange" specifies Lagrange polynomials and "2" is the polynomial degree (i.e., quadratic polynomials).

Recall that Laplace's equation is

$$\nabla^2 u = \frac{d^2 u}{dx^2} + \frac{d^2 u}{dy^2} = f(x, y) \tag{14.1}$$

and the same boundary conditions as before will be used here ($u = \sin(\pi y)$ on the west boundary, $u = 0$ elsewhere). We begin by integrating both sides of this equation over the entire domain and multiplying both sides of the equation by a "test" function, v, giving

$$\int_\Omega (\nabla^2 u)(v) \quad d\Omega = \int_\Omega f(x, y) \cdot v \, d\Omega. \tag{14.2}$$

Using integration by parts, this equation becomes

$$\int_\Omega (\nabla u)(\nabla v) \, d\Omega + \text{boundary terms} = \int_\Omega f(x, y) \cdot v \, d\Omega. \tag{14.3}$$

This form of the equation is called the *weak form*, and the boundary terms cancel along all interior boundaries. Along external boundaries, the boundary terms are used to enforce Neumann boundary conditions or they are not present when we have Dirichlet boundary conditions.

14.3.3 Wait! Why Are We Doing This?

The short answer is to ask a mathematician. The long answer will result when you ask a mathematician. The abbreviated answer is that we are trying to find a polynomial approximation of u on every element. Between elements the approximate solution is continuous, but the derivative is not continuous. By multiplying by a test function, we were able to move one of the two derivatives off of u and onto v. Now we have an equation that we can actually evaluate over the domain to determine an approximate solution. On each element we are trying to determine, essentially, the polynomial coefficients for that element. Since we have many elements, the result is a large linear system where the unknowns are the polynomial coefficients for all the elements.

14.3.4 FEniCS Implementation

The Python script below uses FEniCS (primarily dolfin) to solve Laplace's equation on the unit square.

```python
from dolfin import *

# Create mesh
mesh = UnitSquareMesh(32, 32)

# Create function space
V = FunctionSpace(mesh, 'Lagrange', 2)

# Define boundary conditions
u0 = Expression('sin(pi*x[1])*(1-x[0])')

def u0_boundary(x, on_boundary):
```

```
    return on_boundary

bc = DirichletBC(V, u0, u0_boundary)

# Define variational problem
u = TrialFunction(V)
v = TestFunction(V)
f = Constant(0.0)
a = inner(grad(u), grad(v))*dx
L = f*v*dx

# Compute solution
u = Function(V)

solve(a == L, u, bc,
      solver_parameters={'linear_solver': 'cg',
                         'preconditioner': 'ilu'})

# Dump solution to file in VTK format
file = File('poisson.pvd')
file << u
```

The construction of the matrix problem happens when the matrix object, *a*, is created. The inner product of "grad(u)" and "grad(v)" is same as the weak form equation above. The right-hand side, *L*, is just a vector of zeros. The dolfin solve() function solves the linear matrix problem using an iterative method (conjugate gradient) with an incomplete matrix factorization as the preconditioner. The final result is written to a file and can be visualized using Paraview, Visit, or other visualization software packages. Figure 14.2 was generated using Visit from Lawrence Livermore National Laboratory.

14.4 Pattern Formation

Have you ever wondered how the hairs on a zebra know whether they should be white or black so that distinct stripes ultimately form? In nature, there are many, many patters, from the spots on a leopard to the observation that octopuses and spiders consistently have eight legs. The underlining biological mechanism that allows for the formation of patters was a mystery until Alan Turing proposed a reaction-diffusion system that allowed for stable patterns to form from random initial noise [2]. The Turing system consists of two diffusible chemicals: an *activator* that diffuses more slowly and over shorter distances and an inhibitor that diffuses more rapidly. Starting with a random initial pattern (Figure 14.3, top left), the activator compound causes more

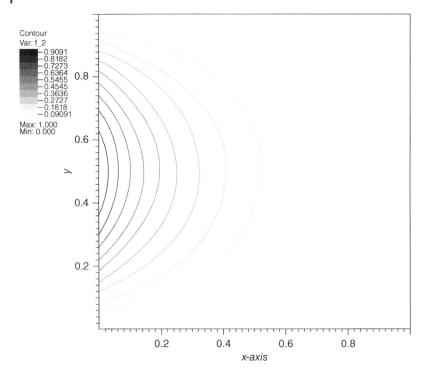

Figure 14.2 FEM solution to the model problem (Laplaces equation) using a second-order Lagrange function space. The solution was visualized using Visit.

activator to be produced. In a positive feedback loop, regions that had slightly higher activator concentrations in the initial, random pattern, these areas quickly achieve even higher activator concentrations (Figure 14.3, top right). However, because the activator moves more slowly, the increases in activator concentration are very local. The inhibitor, on the other hand, moves over greater distances and the regions between the high activator peaks quickly become suppressed by the inhibitor (Figure 14.3, lower right). The result is a stable, predictable pattern of peaks and values. The number of peaks depends on the relative diffusion or movement rates of the two molecules and the speed of the activation and inhibition reactions.

One system of equations that describes the reaction-diffusion system of Turing and others is

$$\frac{\partial A}{\partial t} = D_A \left(\frac{\partial^2 A}{\partial x^2} + \frac{\partial^2 A}{\partial y^2} \right) + A^2 \cdot H - A \tag{14.4}$$

$$\frac{\partial H}{\partial t} = D_H \left(\frac{\partial^2 H}{\partial x^2} + \frac{\partial^2 H}{\partial y^2} \right) - A^2 \cdot H + 1, \tag{14.5}$$

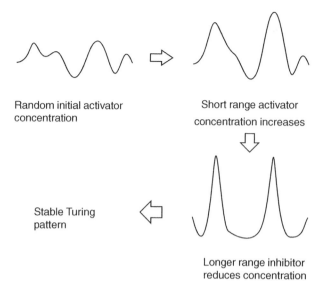

Random initial activator concentration

Short range activator concentration increases

Longer range inhibitor reduces concentration between peaks

Stable Turing pattern

Figure 14.3 Pattern formation based on the reaction-diffusion system of Turing begins with a random concentration of activator (shown in upper left) and inhibitor (opposite of activator). Regions of greater initial activator concentration start to grow due to a positive feedback loop (upper right). The faster moving inhibitor suppresses the activator between the initial peaks (lower right) and ultimately leads to a stable pattern of peaks and valleys.

where A is the activator concentration, H the inhibitor concentration, and D the diffusivity of the two different species. Equations 14.4 and 14.5 are partial differential equations in both space and time, but the greatest challenge is the $A^2 \cdot H$ term, which makes the equations nonlinear. There are a few different approaches for solving nonlinear equations in FEniCS, and the script below utilizes one approach for this system of equations.

```
from dolfin import *
from numpy.random import random
import numpy

class TuringPattern(NonlinearProblem):
    def __init__(self, a, L):
        NonlinearProblem.__init__(self)
        self.L = L
        self.a = a
    def F(self,b, x):
        assemble(self.L, tensor=b)
    def J(self, A, x):
        assemble(self.a, tensor=A)
```

```
# Load mesh from file
mesh = UnitSquareMesh(48,48)

# Define function spaces (P2-P1)
U = FunctionSpace(mesh, "CG", 2)
W = U * U

# Define trial and test functions
du = TrialFunction(W)
(q, p) = TestFunctions(W)

# Define functions
w = Function(W)
w0 = Function(W)

# Split mixed functions
(dact, dhib) = split(du)
(act, hib) = split(w)
(act0, hib0) = split(w0)

# Set parameter values
dt = 0.05
T = 20.0

# Initial conditions
class IC(Expression):
    def eval(self,values,x):
        values[0] = 1.0*random() + 0.25
        values[1] = 1.0*random() + 0.25
    def value_shape(self):
        return(2,)

w_init = IC(element=W.ufl_element());
w.interpolate(w_init)
w0.interpolate(w_init)

L0 = act*q*dx - act0*q*dx \
    + dt*0.0005*inner(grad(act), grad(q))*dx \
    - dt*inner(act*act*hib,q)*dx \
    + 1.0*dt*inner(act,q)*dx
L1 = hib*p*dx - hib0*p*dx \
    + dt*0.1*inner(grad(hib), grad(p))*dx \
    + dt*inner(act*act*hib,p)*dx \
```

```
      - dt*inner(Constant(1.0),p)*dx
L = L0 + L1
a = derivative(L, w, du)

# Create files for storing solution
ufile = File("results/pattern.pvd")

# Create nonlinear problem and Newton solver
problem = TuringPattern(a, L)
solver = NewtonSolver()

# Time-stepping
t = dt
while t < T + DOLFIN_EPS:
    print "t =", t
    w0.vector()[:] = w.vector()
    solver.solve(problem, w.vector())

    # Save to file
    ufile << w.split()[0]

    # Move to next time step
    t += dt
```

Most of the functions in this script are the same as those previously described when Laplace's equation was solved. The TuringPattern object at the top is derived from built in functionality in FEniCS for automatically solving nonlinear problems using a Newton solver. At each time step, the activator concentration is saved to a file. Figure 14.4 shows the pattern that results from the script above. By changing the relative sizes of the various terms in equations 14.4 and 14.5 , different patterns (or no pattern at all) can be achieved.

Additional Resources

The following books may be useful for learning more about the FEM.

- An Introduction to the Finite Element Method by Reddy [3]
- Finite Element Methods for Flow Problems by Donea and Huerta [4]
- The Finite Element Method in Heat Transfer and Fluid Dynamics by Reddy and Gartling [5]
- The Mathematical Theory of Finite Element Methods by Brenner and Scott [6]

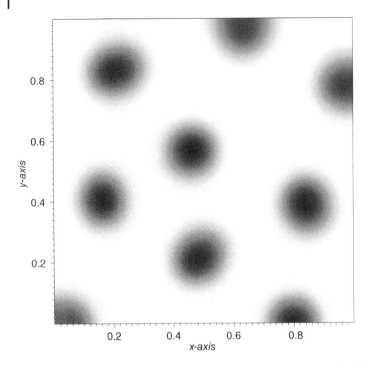

Figure 14.4 Activator concentration resulting from a Turing pattern simulation.

References

1 Logg, A., Mardal, K.A., and Wells, G.N. (2012) *Automated Solution of Differential Equations by the Finite Element Method*, Springer-Verlag, Berlin Heidelberg, doi: 10.1007/978-3-642-23099-8.

2 Turing, A. (1952) The chemical basis of morphogenesis. *Philos. Trans. R. Soc. London, Ser. B*, **237**, 37–72.

3 Reddy, J. (1993) *An Introduction to the Finite Element Method*, McGraw Hill, Boston, MA, 2nd edn.

4 Donea, J. and Huerta, A. (2003) *Finite Element Methods for Flow Problems*, John Wiley & Sons, Ltd., West Sussex, England.

5 Reddy, J. and Gartling, D. (1994) *The Finite Element Method in Heat Transfer and Fluid Dynamics*, CRC Press, Boca Raton, FL.

6 Brenner, S. and Scott, L. (2002) *The Mathematical Theory of Finite Element Methods*, Springer-Verlag, New York, NY, 2nd edn.

Index

Chemical and Biomedical Engineering Calculations Using Python®, First Edition. Jeffrey J. Heys.
© 2017 John Wiley & Sons, Inc. Published 2017 by John Wiley & Sons, Inc.
Companion Website: www.wiley.com/go/heys/engineeringcalculations_python

Printed and bound by CPI Group (UK) Ltd, Croydon, CR0 4YY

27/10/2024

14580277-0001